Autorenkollektiv

Organic Chemistry

Autorenkollektiv

ORGANIC CHEMISTRY

With 12 Figures and 59 Tables

AKADEMIE-VERLAG · BERLIN
1982

Die Originalausgabe erscheint im Springer-Verlag Berlin · Heidelberg · New York

Vertrieb ausschließlich für die DDR und die sozialistischen Länder

Erschienen im Akademie-Verlag, DDR-1086 Berlin, Leipziger Straße 3—4
Lizenznummer: 202 · 100/538/82
Gesamtherstellung: VEB Druckerei „Thomas Müntzer“, 5820 Bad Langensalza
Umschlaggestaltung: Karl Salzbrunn
Bestellnummer: 763 072 0 (6682) · LSV 1215
Printed in GDR
DDR 92,— M

Table of Contents

Radical Eliminations from Gaseous Cation Radicals via Multistep Pathways — The Concept of "Hidden" Hydrogen Rearrangements

Prof. Dr. Helmut Schwarz*)

Institut für Organische Chemie der Technischen Universität Berlin, Berlin (FRG)

Table of Contents

---

* Presented in part as a plenary lecture at the Euchem Conference on "The Chemistry of Ion Beams", Lunteren, 1980.

1 Introduction

The eliminations of neutral closed-shell species HX from cation radicals $R - X]^{+\cdot}$ are well known in mass spectrometry. They have been studied in detail with regard to the reaction mechanisms underlying the hydrogen rearrangement, the ion structures formed upon dissociation and the energetics of the reactions[1]. Representative examples of this type of unimolecular processes of gaseous ions are, for instance, the loss of HX *via* the operation of an *ortho* effect *1*→*2*[2] or alkene elimination from ionized carbonyl compounds *via* the McLafferty rearrangement *3*→*4*→*5* (1)[3].

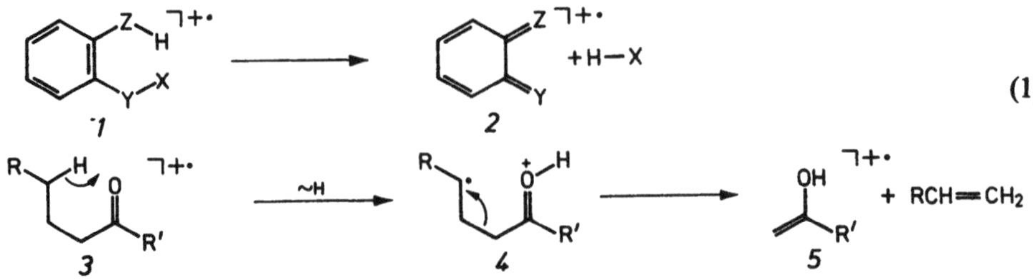

(1)

A common feature of the reactions described in (1) is that the hydrogen migration can be traced directly from appropriate mass shifts in the spectra of suitably ^2H labelled precursor ions.

Rearrangements must be involved also in radical eliminations in which the open shell species $X^{\cdot}$ is not present as a structural unity of $M^{+\cdot}$, but has to be created prior to the eventual dissociation step. Examples of such processes are the loss of $H\dot{C} = O$ from ionized methylvinylsulfide *6*→*7*[4] or the elimination of both $OH^{\cdot}$ and $H\dot{C} = O$ from the cation radical of *8*[5,6] (the arrows in (2) indicate the atoms of *8* which are involved in the formation of the radicals $OH^{\cdot}$ and $HCO^{\cdot}$. Both radicals are obviously not present as structural functions in the precursor. For $HCO^{\cdot}$ loss from *6* no labelling work has been reported).

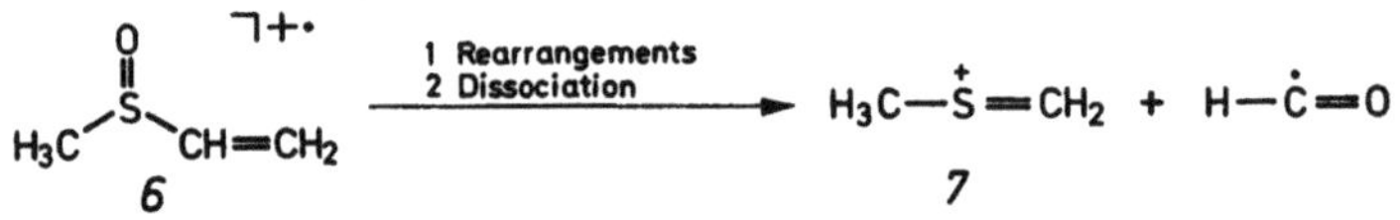

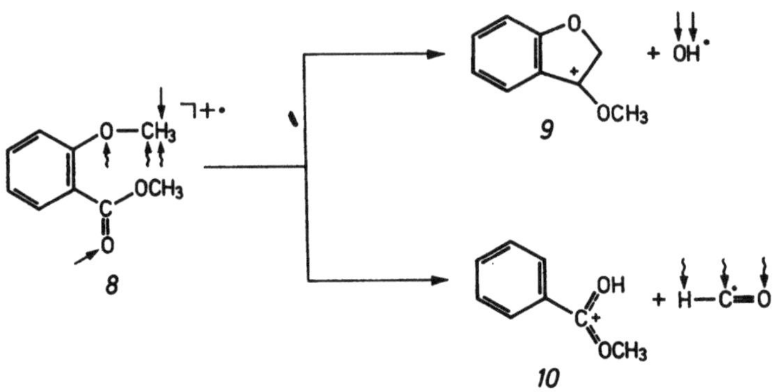

(2)

Essentially different from the reactions described in (2) are radical eliminations in which the radical $X^{\cdot}$ is already present in the ionized molecule $R_1 - X^{]+\cdot}$, but where the actual dissociation of the $R_1 - X$ bond is preceded or accompanied by an isomerization of the charge carrying part R. This situation (3) may occur in cases, in which the direct cleavage *11→12* is energetically less likely than the two-step (or generally: multi-step) pathway *11→13→14*.

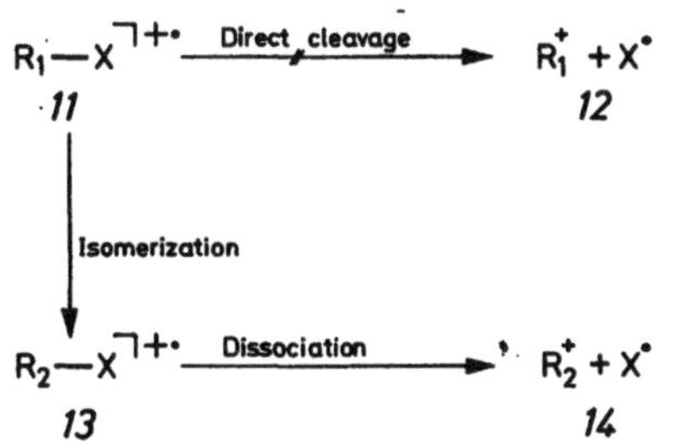

(3)

The multistep radical elimination may involve the generation of discrete intermediates, which for instance could be formed by a cyclization process[7] such as *15→16→17*[8]. Alternatively, there may be no intermediate involved in the elimination sequence, but the actual transition states *19*, *22* are substantially lower in energy due to the anchimeric assistance of suitable functional groups[9,10] (4).

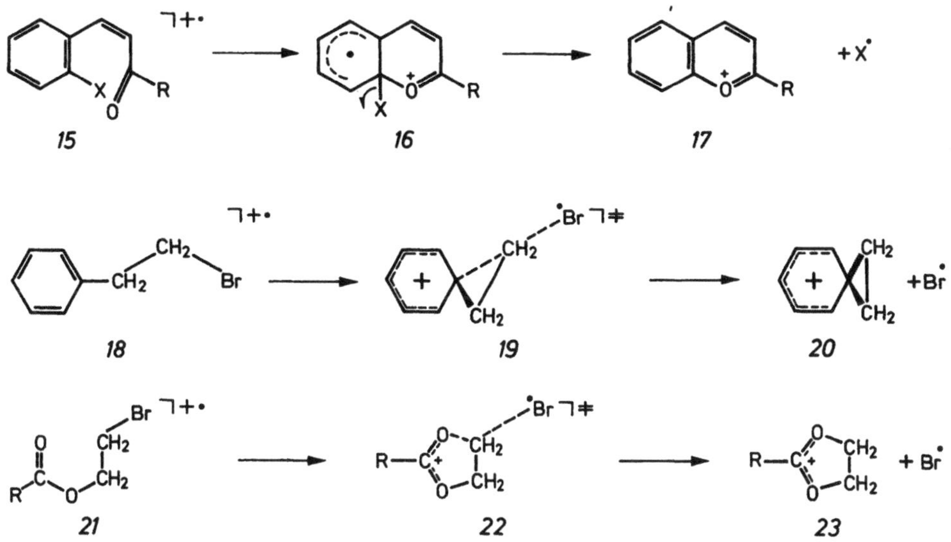

(4)

There is another mechanistic variation to circumvent the direct cleavage process *11→12*. The isomerization *11→13* does not necessarily have to involve a skeletal reorganization. For example, it can be envisaged that a specific hydrogen transfer occurs onto a suitable acceptor function *Y*, *24→25* (5). In this way a reactive radical site is created which induces the dissociation step (elimination of $X^{\cdot}$) by simultaneous double bond formation. Note that in that case the *hydrogen migration takes place within the charge carrying part of the molecule* $R - \dot{X}^{]+\cdot}$ *and that the migrated hydrogen atom remains there throughout the whole process so that the intervention of a H-migration cannot be detected directly by appropriate mass shifts in the spectra of* ^2H *labelled precursors.* However, the mechanistic nature of such a "hidden" hydrogen rearrangement can be revealed by several other methods like the measurement of primary kinetic isotope effects, the investigation of stereoisomeric precursors, the study of the energetics of the process and the structure determination of the ionic product *26* formed in reaction (5).

It is the purpose of this review[11] to show that a fairly substantial number of quite different unimolecular dissociations of gaseous cation radicals can be described conveniently by this concept. Among the reactions discussed are some unusual cleavage processes, the occurrence of which were difficult to explain without the involvement of "hidden" hydrogen migrations.

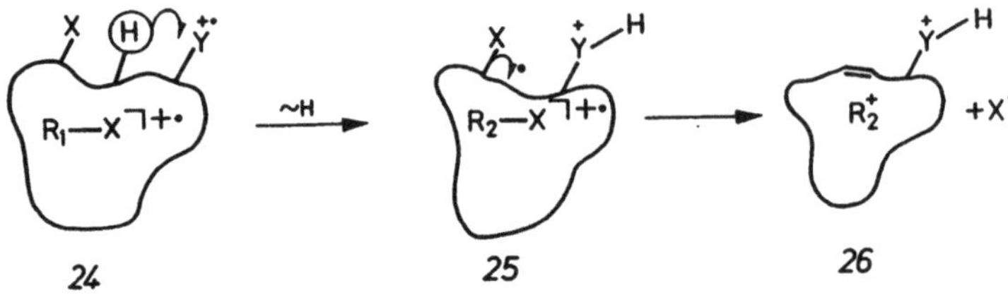

(5)

2 Specific (C—C) Bond Cleavage Induced by Hydrogen Migration to a Carbonyl Function

The electron impact mass spectrum of methyl isobutyrate, *27*, contains a signal for a $[M-CH_3]^+$ ion. ^2H- and ^{13}C-labelling clearly establishes that the eliminated methyl radical originates exclusively from the intact $^{\beta}CH_3$ group (6)[12,13,14].

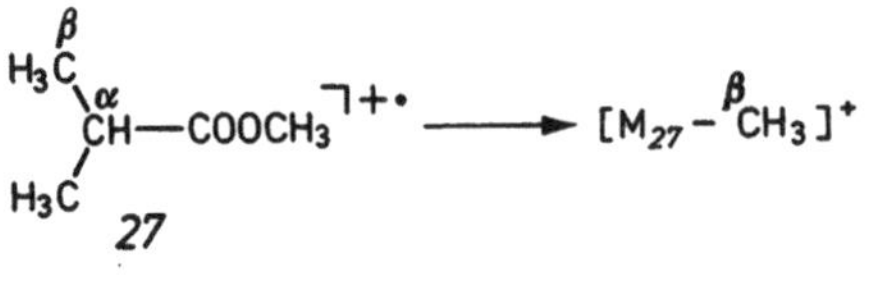

(6)

However, there are several experimental observations which are against an interpretation of the reaction as being a one-step C—C cleavage process which would give rise to the formation of the relatively unstable α-oxocarbenium ion *28*.

1) Under field ionization kinetic conditions (FIK) the process of $CH_3^{\cdot}$ elimination from ionized *27* is not observed at $t < 10^{-10}$ sec. This is a surprising result bearing in

mind that the period of a bond vibration is of much shorter time (t $\approx 10^{-13}$ sec.) and that direct one-bond cleavage reactions were expected to occur at a time substantially shorter than 10^{-10} sec. Moreover, the relative rate of $[M-CH_3]^+$ ion formation as a function of the molecular ions lifetime shows a pronounced maximum at 10^{-8} sec. In general, such a maximum indicates the occurrence of a multistep pathway[15]. 2) The relative abundance of the $[M-CH_3]^+$ ion current is *increasing* with *decreasing* energy of the ionizing electrons. In particular, for metastable ion decompositions the $[M-CH_3]^+$ ion currents are about 70% of the total ion current (second field-free region). This observation is typical for a reaction requiring both a relatively low activation energy and a highly specific arrangement of atoms in the transition state such as in a rearrangement[16,17]. 3) The collisional activation (CA)[18] spectrum of the $[M-CH_3]^+$ ions formed from *27* is identical with that of the $[MH]^+$ ions of methylacrylate *30* formed upon chemical ionization[19] with the gaseous Brønsted acids CH_5^+ and H_3^+. It is not unlikely that protonation of *30* will occur at the carbonyl oxygen atom, thus giving rise to *29*. From the identical CA spectra of both isobaric ions it may be concluded that a common structure is formed and, consequently, that the $[M-CH_3]^+$ ions from *27* can be represented by structure *29*. Moreover, the generation of the abundant fragment ion *31* at m/z *27* ($C_2H_3^+$) upon collisional activation is diagnostic for *29*, but not for *28*. 4) From the ^2H labelled ester *27a* the radicals $CD_3^{\cdot}$ and $CH_3^{\cdot}$ are eliminated in a ratio 3.2:1 at molecular ion lifetimes of about 10^{-6} sec. This result is neither compatible with the direct C—C bond cleavage *27*→*28* nor with the operation of a secondary isotope effect.

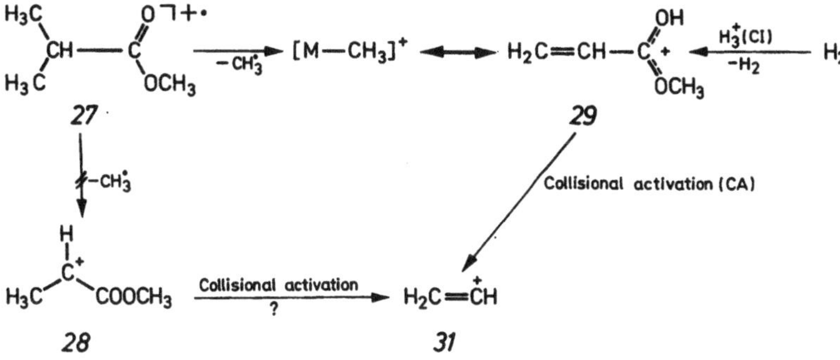

(7)

However, all results (7) and (8) are consistent with a *two*-step process, in which the reaction commences with a hydrogen migration from one of the $^{\beta}CH_3$ groups to the ester function. The dissociation is completed with a radical induced C—C cleavage, thus eliminating the other (still intact) $^{\beta}CH_3$ group and giving rise to the formation of protonated methyl acrylate. The fact that *27a* eliminates $CD_3^{\cdot}$ in favour of CH_3 is a direct consequence of the two-step nature of the reaction and indicates that the hydrogen transfer is the rate determining step. Thus, *27a* rearranges more easily to *32a* (H transfer) than to *32b* (D transfer) due to the operation of a primary isotope effect with the consequence that $CD_3^{\cdot}$ loss is more enhanced than elimination of $CH_3^{\cdot}$. The ratio of $[M-CD_3]^+/[M-CH_3]^+$ reflects directly the magnitude of the primary kinetic isotope effect operating in the "hidden" hydrogen/deuterium migration (8).

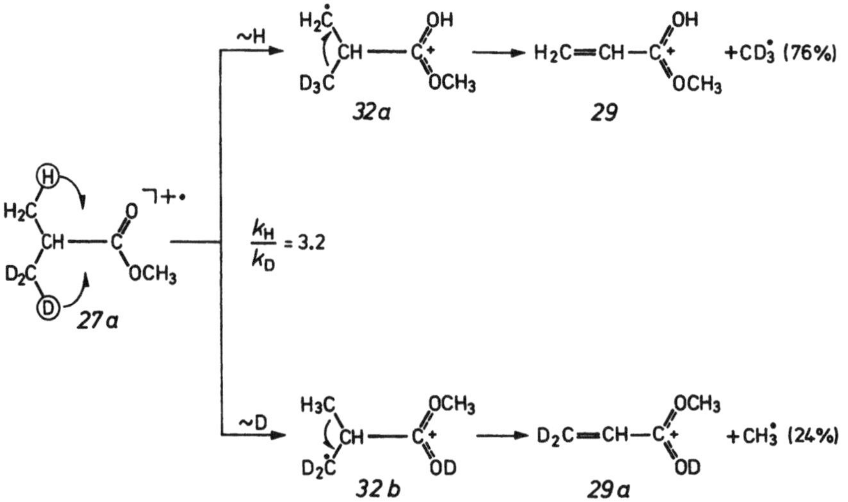

(8)

Similarly, it is observed that ionized 2-methylbutanoic acid **33** loses $CH_3^{\cdot}$ more readily than $C_2H_5^{\cdot}$ [14]. Again, the ion abundance ratio $[M-CH_3]^+/[M-C_2H_5]^+ = 3.5$ is best compatible with a two-step mechanism in which the intermediate *34* (*36*) is formed in a rate-determining hydrogen transfer. It is well-known that the migratory aptitude of a "secondary" hydrogen is better than that of a "primary" hydrogen, so that the intermediate *34* is formed preferentially and, thus, more $CH_3^{\cdot}$ loss can be expected and *vice versa* (9).

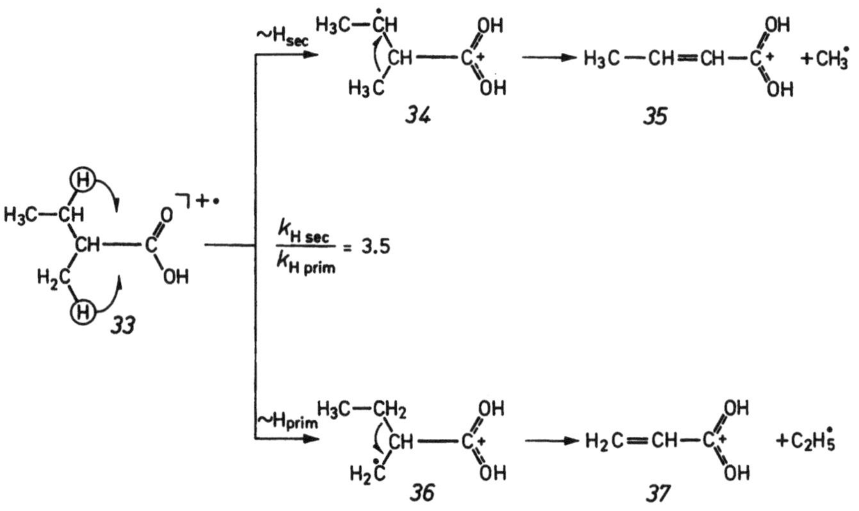

(9)

The elimination of the original $CH_3^{\cdot}$ group from ionized *n*-butanoic acid, *38*, was shown to proceed also mainly *via* a two-step process. The reaction is initiated by a hydrogen migration *via* a four-membered transition state *38*→*39*, which appears to arise from an isolated electronic state of the molecular ion of *38*. From the intermediate *39* $CH_3^{\bullet}$ is eliminated in close analogy to the reactions described above[20] (10).

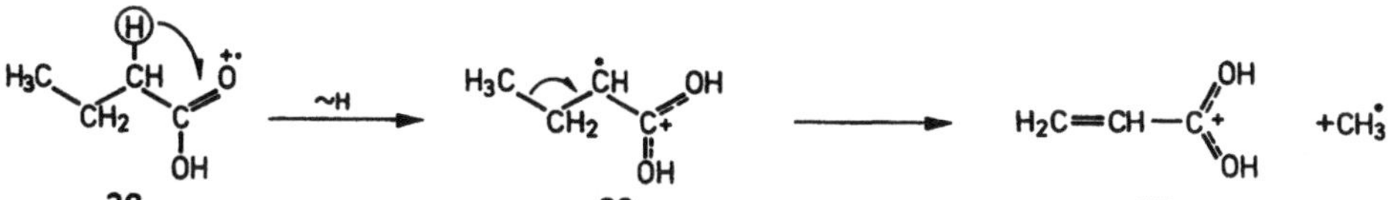

38 39 37

(10)

The long-standing controversy whether the cleavage of the ($^{\delta}$C—R) bond in ionized *40* (R = alkyl, halogen) occurs *via* an S_Ri reaction *40*→*41*[21] or *via* a two-step process[22], commencing with the $^{\gamma}$H-transfer of the allylically activated hydrogen *40*→*42* and terminating with the radical induced dissociation *42*→*43* was solved unequivocally by CA spectroscopy[23] (11). It was demonstrated, that 1) *41* and *43*, generated independently from ionized *44* and *45* *via* direct α-cleavage, do not interconvert under CA conditions and that 2) the CA spectrum of the $[M—R]^+$ ions formed from *40* is identical with that of the ions *43* generated from *45*, whereas the CA spectrum of the $[M—Y)^+$ ions from *44* exhibits diagnostically useful differences. Thus, the direct S_Ri type process *40*→*41* can be ruled out with certainty in favour of the two-step reaction *40*→*42*→*43*.

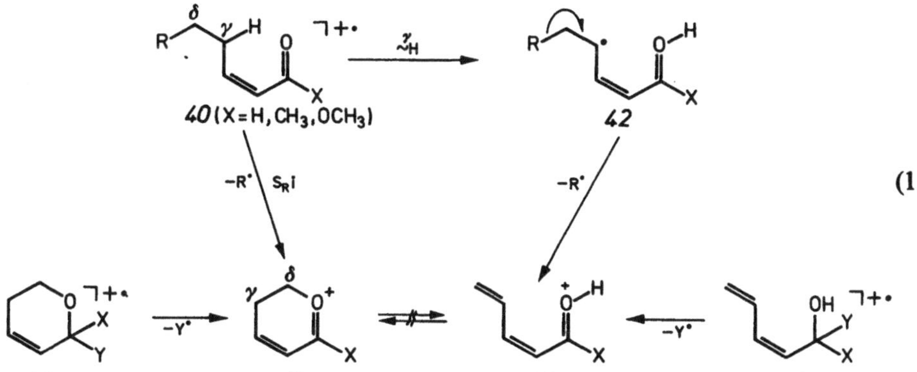

(11)

The fact that from the ionized ester *46* mainly $C_2H_5^{\bullet}$ is eliminated whereas the stereoisomeric compound *47* decomposes preferentially *via* $CH_3^{\bullet}$ loss[24] is best explained[25] by the operation of a two-step process (12). The regiospecific hydrogen transfer *46*→*48* is responsible for $C_2H_5^{\bullet}$ loss, whereas $CH_3^{\bullet}$ is likely to be eliminated from the intermediate *50*. It should be mentioned that the different reactivities of the molecular ions of *46* and *47* indicate that no stereomutation of the double bond occurs prior to the hydrogen transfer.

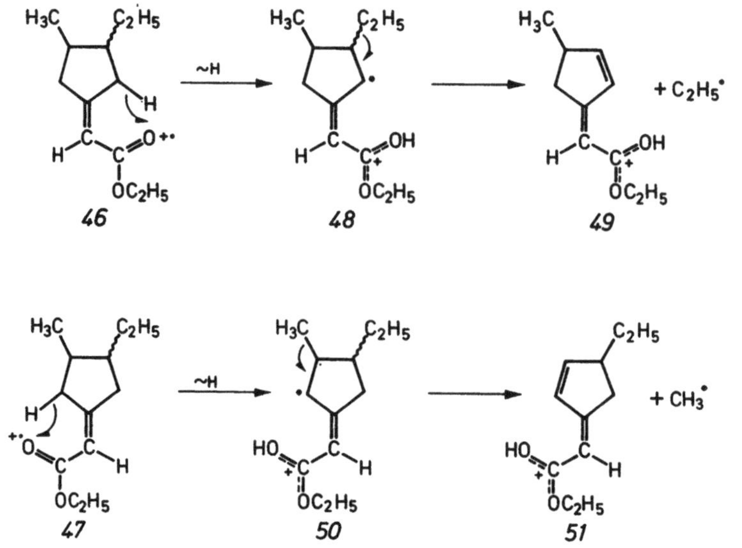

(12)

Methyl loss from the ionized α, β unsaturated ester *52* constitutes a problem, the mechanism of which is quite intriguing. ^2H labelling experiments[26] clearly show that the eliminated methyl radical originates exclusively from the methyl group attached to the β carbon of *52*. However, the assumption of a direct C—C cleavage reaction which would give rise to the formation of an extremely unstable vinyl cation *53* (13) is not in accordance with the observation of an abundant metastable ion signal which itself points to a rearrangement process. There are two alternatives which in principal can account for this unusual C—C bond cleavage[26,27]. 1) $CH_3^{\cdot}$ loss from the $^{\beta}$carbon atom is assisted by an intromolecular "Michael addition" type reaction in which the transition state energy for the cleavage of the C—C bond is lowered by interaction of the incipient vinyl cation center with the migrating methoxy function eventually giving rise to a presumably stable acylium ion *55*; this ion is able to delocalize its charge substantially. 2) As an alternative to the "Michael addition", the multistep pathway *52*→*56*→*57*→*58* has been invoked. This sequence commences with a hydrogen migration within the methyl ester function (*52*→*56*) and then the $^{\beta}CH_3^{\cdot}$ group is eliminated either after cyclization to *57* or directly *56*→*58*. All experiments carried out so far (i.e. investigation of ^2H labelled precursors, study of stereoisomeric pairs of *52*, and the determination of the kinetic energy release T) cannot distinguish between the two fundamentally different mechanism described in (13). However, a recent reinvestigation of the problem[28] employing CA spectroscopy as a tool for an unequivocal structure elucidation of the $[M-CH_3]^+$ ions formed from *52* clearly indicates that a major part of the fragment ion has structure *58* and not *55*. Thus, the unusual $^{\beta}CH_3^{\cdot}$ elimination from the molecular ion of *52* can be viewed as a further example for the operation of "hidden" hydrogen migration preceding C—C bond cleavage.

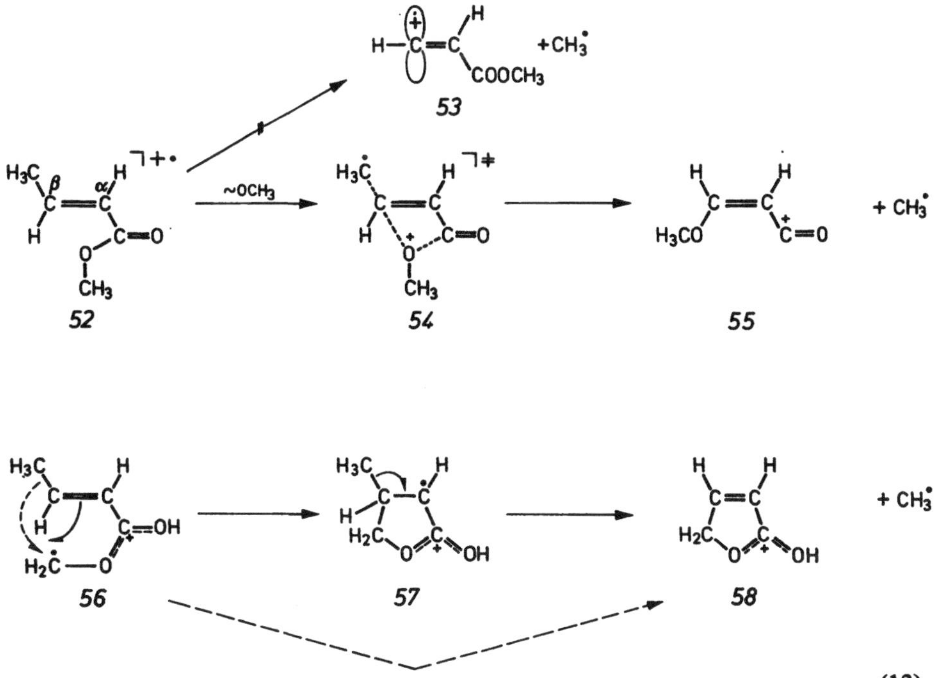

(13)

3 "Saponification" of Ionized Esters by Intramolecular Hydrogen Transfer: Cleavage of the Alkyl/Oxygen Bond

The alkyl cleavage in ionized esters *59* according to (14) is not favoured due to the fact that the incipient ion *60* is not stable at all, whereas ion *61* and *62* are calculated to be relatively high in energy[29,30,31]. Thus, the formation of these species from simple aliphatic esters has never been achieved[32]. Ions of the elemental composition RCO_2^+, which are found experimentally to be structurally different from isomeric oxonium ions, could only be generated by charge exchange experiments[30] presumably as *62* from the corresponding carboxylate anions RCO_2^- *63*.

However, several examples of reactions in which a formal cleavage of the alkyl/oxygen bond in esters occurs, have been presented recently[2,5,6]. A study of suitable precursors *64* in combination with kinetic isotope effect measurements, kinetic energy release determination and structure elucidation of the resulting fragment ion by CA spectroscopy has revealed that the cleavage of the alkyl/oxygen bond in *64* is associated with the formation of the substituted carboxylic acid *66* (15). The fact that the *meta/para* isomers of *64* do not eliminate a methyl radical from the ester function points to the operation of an *ortho* effect, which obviously is essential for the process.

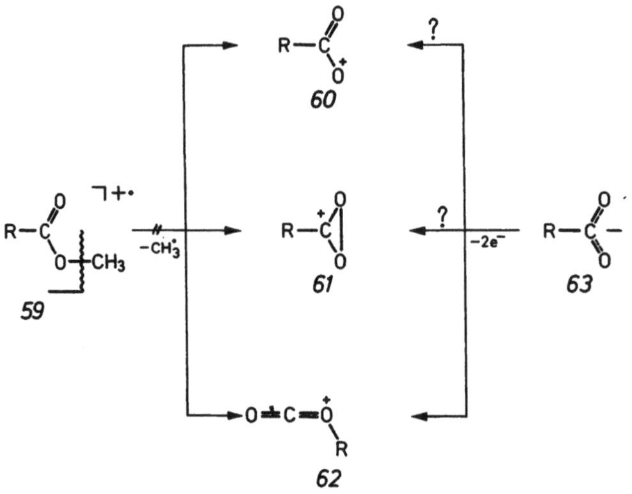

(14)

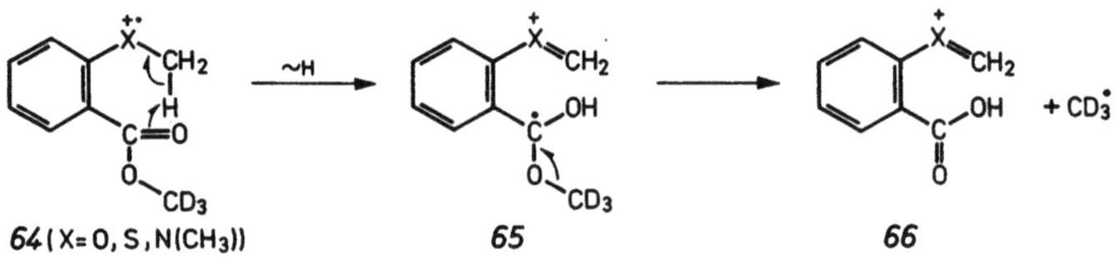

(15)

A hydrogen migration, operative *via* an *ortho* effect, is also involved in the methyl loss from ionized *67*[6] (16). ^2H labelling demonstrates that the eliminated methyl radical originates from the ester function and not from the methyl group attached directly to the benzene ring. From CA it has to be concluded that the $[M-CH_3]^+$ ion has a structure identical with that formed upon $Br^\cdot$ loss from ionized *70*. However, the CA spectrum of the isomeric ion *73*, formed upon protonation of *74*, is substantially different from that of *69*. Thus, the multistep pathway *67→73* commencing with a hydrogen migration within the ester function *67→71* can be discarded in favour of a process in which a hydrogen atom from the benzylic site is initially transferred to the carbonyl oxygen *67→68*.

With regard to the spatial requirement of the hydrogen migration involved in "saponification" reactions and in particular the question concerning the geometrical distance between the hydrogen donor and the acceptor functions, some details could be derived from the investigation of suitably disubstituted furans *75*, *76* and *77*[33]. Inspection of Dreiding models of the neutral molecules indicated that the distance between the CH_3 groups and the carbonyl oxygen is approximately 1.8 Å for both 1.2-disubstituted furans *75* and *76*, but substantially larger for the 3.4-disubstituted isomer *77* (≧2.8 Å). Moreover, electron impact ionization of *75*, *76* and the heterocycles *78*[34,35,36] gives rise to abundant signals for $[M-CD_3]^+$

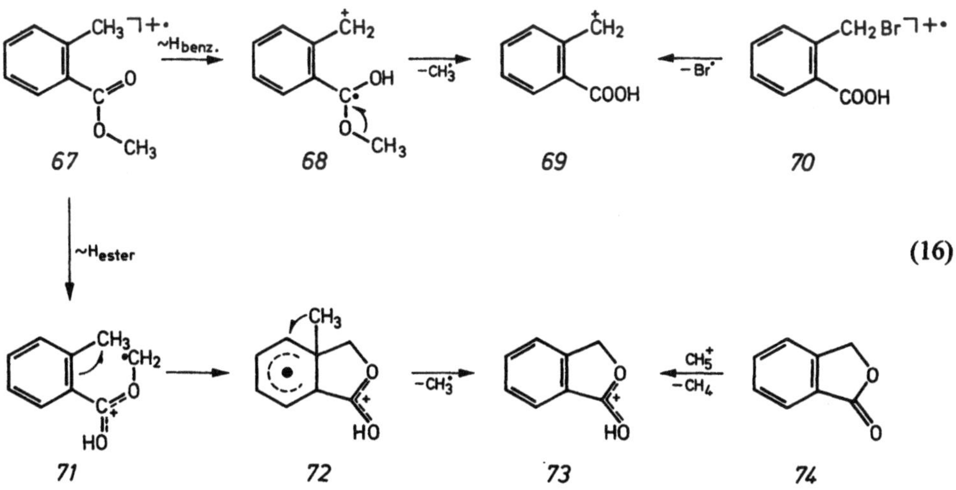
67
~H benz.
68
$-CH_3^{\bullet}$
69
$-Br^{\bullet}$
70
~H ester
71
72
$-CH_3^{\bullet}$
73
CH_5^+
$-CH_4$
74

(16)

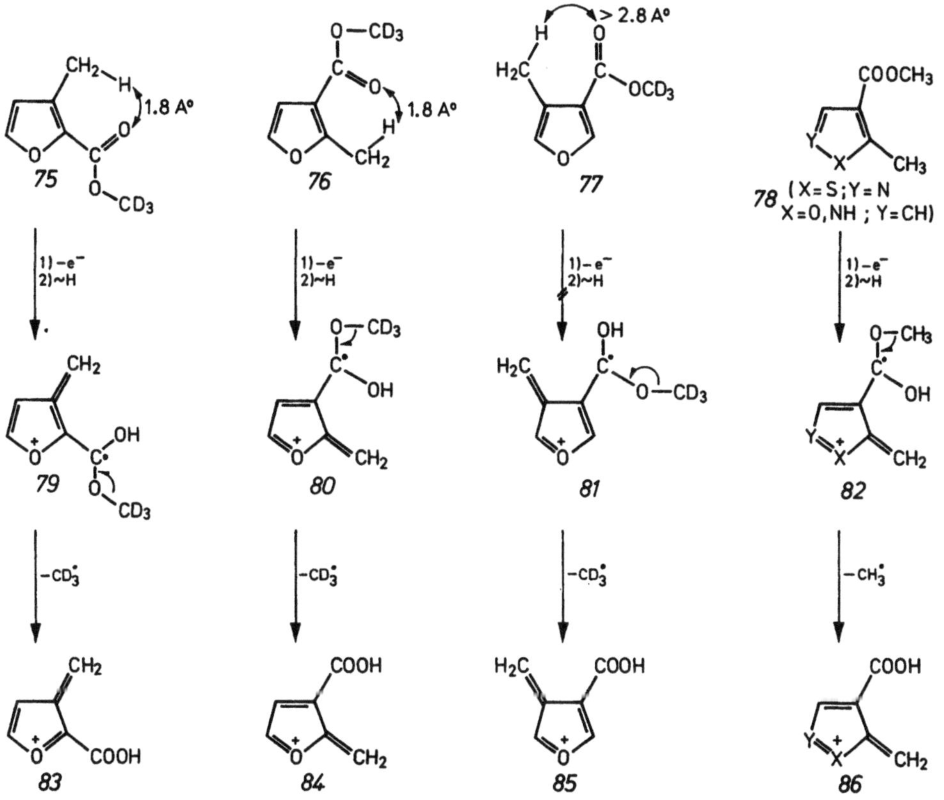
1.8 A°
75
1.8 A°
76
> 2.8 A°
77
78 (X=S; Y=N
X=O,NH ; Y=CH)
1) $-e^-$
2) ~H
79
80
81
82
$-CD_3^{\bullet}$
$-CH_3^{\bullet}$
83
84
85
86

(17)

resp. $[M-CH_3]^+$ (from *78*), but no loss of either CH_3 or $CD_3^{\cdot}$ is observed from ionized *77*[33] (17). We consider these results to provide evidence that the initial hydrogen migration, which must be viewed as a condition necessary for the cleavage of the oxygen/alkyl bond in esters is dependent upon the distance between the two interacting functional groups as strongly as is the case for the McLafferty rearrangement. For the latter a distance of about 1.8 Å is believed to represent an upper limit[3].

4 Cleavage of (O—C) Bonds with Charge Retention on the Oxygen Atom Carrying Part

In close analogy with the non-occurrence of reaction *59→60*, *61* or *62* (14) the cleavage of an oxygen-carbon bond with charge retention on the oxygen atom *87→88* is energetically very unfavourable and, thus, not often — if at all — observed. However, oxenium-ions *88* can be generated quite conveniently by a charge exchange experiment, in which the corresponding alkoxide ion $R'O^-$ *89* is forced to lose two electrons in a special ion/molecule reaction experiment[37] (18).

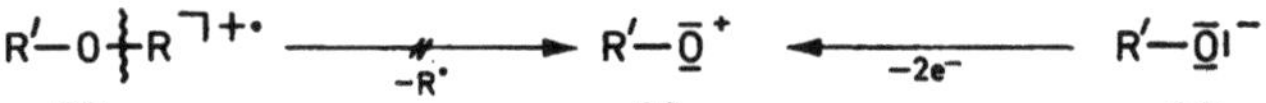

87 *88* *89* (18)

Recently, several examples have emerged which demonstrate that the cleavage of the oxygen-carbon bond in ionized ethers must have occurred. However, a careful examination of the underlying mechanisms indicates, that in all cases so far studied "hidden" hydrogen migrations are involved. For example, the specific elimination of $C_2H_5^{\cdot}$ from ethoxy-substituted benzyl amine is restricted to the *ortho*-isomer *90*[38]. Moreover, from the study of ^2H-labelled precursors and the analysis of the CA spectra[39] is must be concluded that $C_2H_5^{\cdot}$ loss from ionized *90* can be described as a two-step process commencing with the transfer of an activated, benzylic hydrogen atom to the ether oxygen, *90→91* (19).

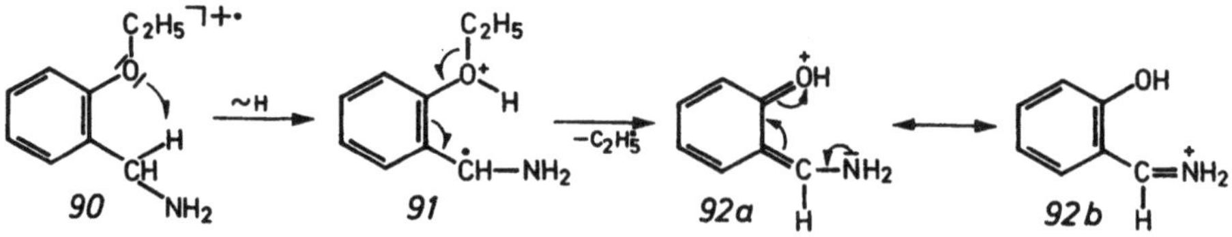

90 *91* *92a* *92b*

(19)

Mechanistically quite different from the process described in (19) is the process of methyl elimination from the cation radicals of *93* and its *meta*/para isomers *96*[40]. In the case of *93* methyl loss is initiated by a hydrogen transfer from the benzylic site to the ester function, *93→94*, thus forming a reactive intermediate *94* from which cleavage of the (O—CH_3) bond is promoted by the formation of a stable carbonyl function *94*→*95* (20).

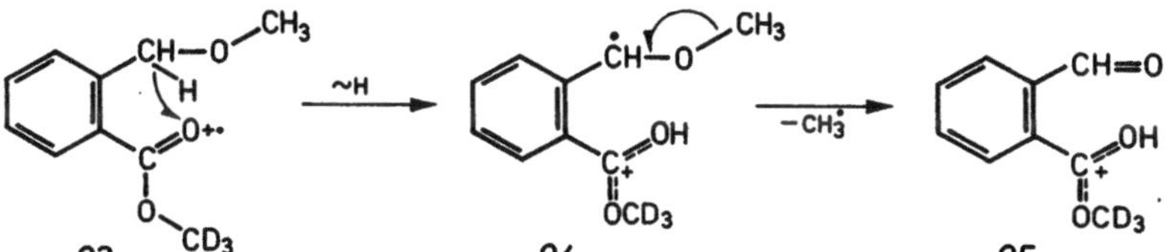

93 *94* *95* (20)

As mentioned above, methyl loss from ionized methoxymethyl derivatives of methyl benzoate is not restricted to the *ortho*-isomer *93*. ^2H labelling has shown that the *meta* and *para* compounds *96* also eliminate $CH_3^{\cdot}$ specifically from the methoxymethyl group. Interestingly, neither from the unsubstituted benzyl methyl ether *97* (X = H) nor from the X-substituted analogues *97* (X = CH_2OH, NO_2) loss of $CH_3^{\cdot}$ is observed (21). Thus, the carbomethoxy function seems to play a decisive role in the dissociation process.

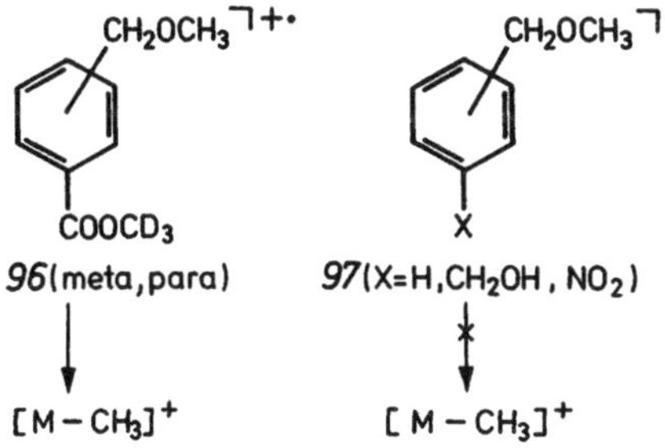

(21)

However, the idea, that *96* may rearrange to the *ortho* isomer *93 via* substituent migration or valence bond tautomerization, which would enable the $CH_3^{\cdot}$ loss to proceed as described in (20), could not be substantiated by experimental facts. For example, the secondary decompositions of the $[M—CH_3]^+$ ions formed from *93* and *96* are different with regard to the reaction channels and both the kinetic energy release and peak shapes associated with the reactions of interest. Moreover, the CA spectra of the $[M—CH_3]^+$ ions exhibit distinct differences. Thus, the $[M—CH_3]^+$ ions posses different ion structures and, consequently, a common intermediate and/or reaction mechanism for the process of methyl elimination from ionized *93* and *96* are very unlikely (22).

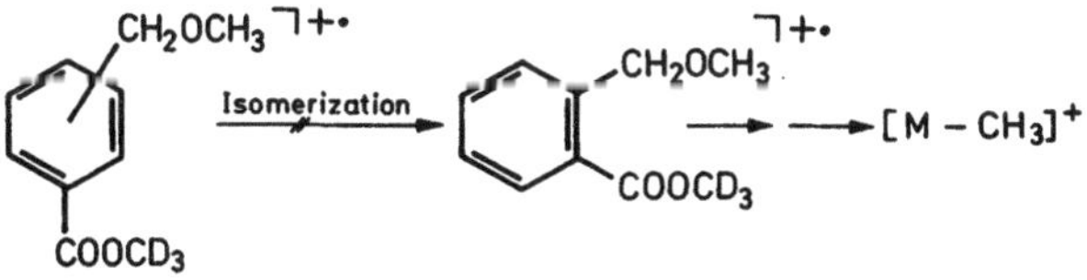

96 *93* (22)

Trans-anular interactions, which would create an active radical site *via* hydrogen transfer through *98*, cannot be invoked to explain the specific loss of a $CH_3^{\cdot}$ radical from the ether side chain. This conclusions is based upon the following experimental observations. The radical cation of the tetrafluoro substituted compound *101* eliminates $CH_3^{\cdot}$, but loss of $CH_3^{\cdot}$ from the *para*-isomer *102* is not observed. If a transanular process according to *97*→*98* were operative, then such a reaction is not expected to be suppressed upon substitution of H by F as is known for many examples from the field of photochemistry of fluoro substituted compounds[41] (23).

Similarly, valence tautomerization is not known to be extremely sensitive to substituent effects. Thus, a fundamentally different reaction mechanism has to be invoked in order to reconcile all experimental facts.

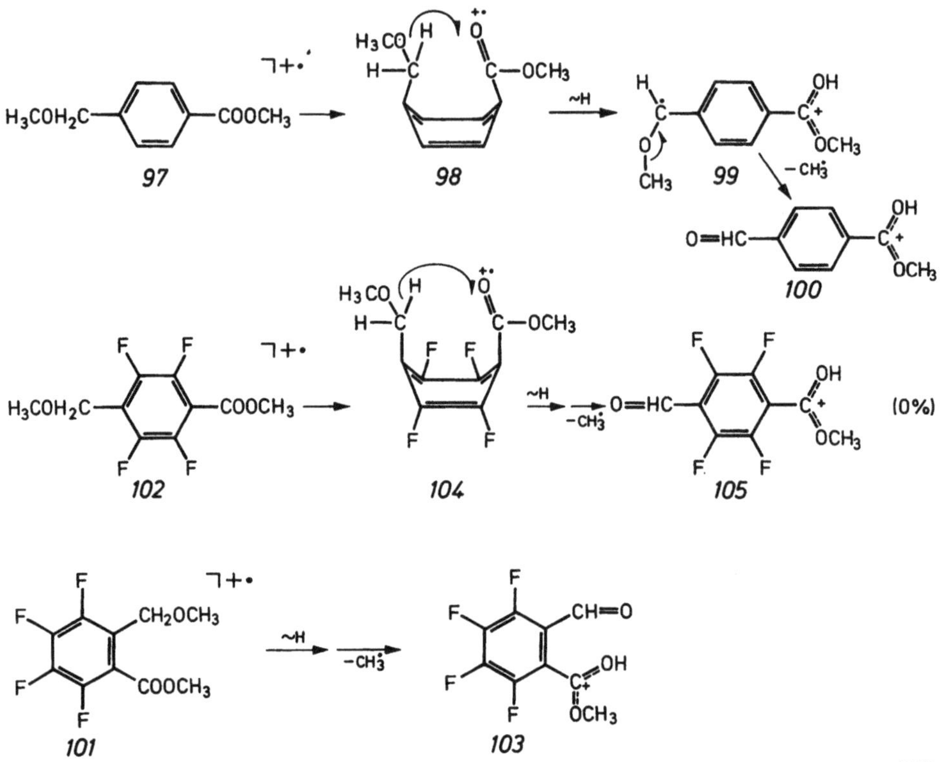

(23)

We suggest that cleavage of the (O—C) bond in ionized *97* commences with a hydrogen transfer from the benzylic position to the benzene ring (24) giving rise to the formation of the intermediate *106*, from which $CH_3^{\cdot}$ is eliminated in a process already described for many other examples. The different reactivity observed for the $[M—CH_3]^+$ ion formed from *93* and *97* is not surprising: Both species are formed with different structures. The fact, that the molecular ion of the tetrafluoro substituted compound *102* is unable to undergo $CH_3^{\cdot}$ loss is not surprising too, because the electron density of the benzene ring in *102* and therefore the ability to form a protonated benzene derivative *106* is likely to be decreased substantially by four fluoride substituents. That a fluoride substituent is capable to reduce the π-electron density of a benzene ring has been shown in the case of intramolecular charge/transfer complexes formedy suitable from substituted p-[2.2]-cyclophanes[42,43].

An alternative to the sequence *97→106→107* has been resently proposed by *Grützmacher*[44]. He suggested that in close analogy to various other well-known processes[45,46,47,48] the $CH_3^{\bullet}$ elimination from ionized *97* commences with a migration of hydrogen from the *ortho*-position to the carbonyl group of the ester function *97→108*. The intermediate *108* may then rearrange to *109* from which the eventual $CH_3^{\bullet}$ loss will take place. Whereas all experimental results discussed so far are compatible with this interpretation, there is one observation which points to the formation of *107* instead of *110*. The $^2H_2$-labelled molecular ions of *97* (labelled in the benzylic methylene group with deuterium) eliminates $CH_3^{\bullet}$, and the resulting $[M-CH_3]^+$ ions undergo loss of both CH_3OH and CH_3OD. This is expected to be the case if *107* in formed, whereas from *110* generated from the same precursor no loss of CH_3OD should occur.

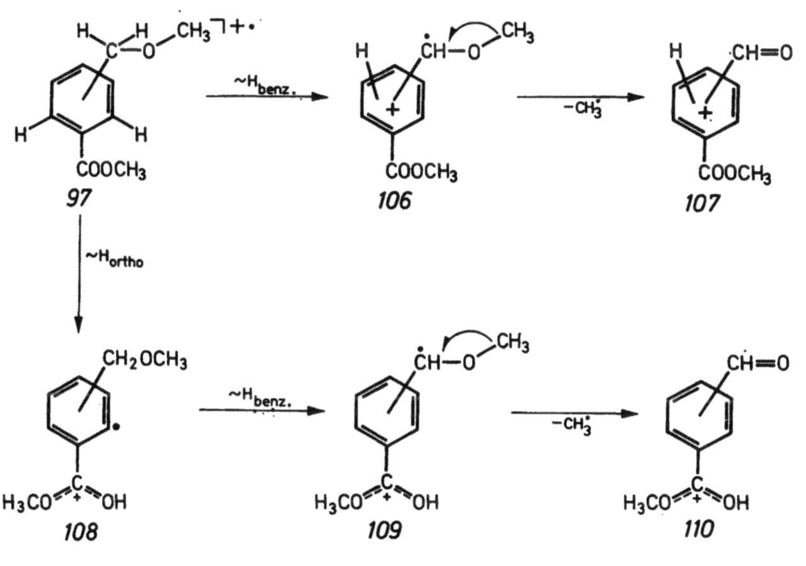

(24)

5 Anchimerically Assisted Dissociations of Cation Radicals with Hydrogen or Trimethylsilicium as the Bridging Structural Elements

The mechanistic details of the cleavage of (S—C) bonds in ionized sulfides, such as *111→113* constitute a challenging problem. Some insight was obtained from a CA spectroscopy study[49] and from the investigation of ^2H-labelled precursors[50,51]. For ionized ethyl methyl sulfide *111* it was concluded from the CA spectra that in addition to the β-fission product *112* the rearranged ion *114* is formed as a stable species in the gas phase. There is no experimental indication for the existence of *113* in a potential minimum (25)[51].

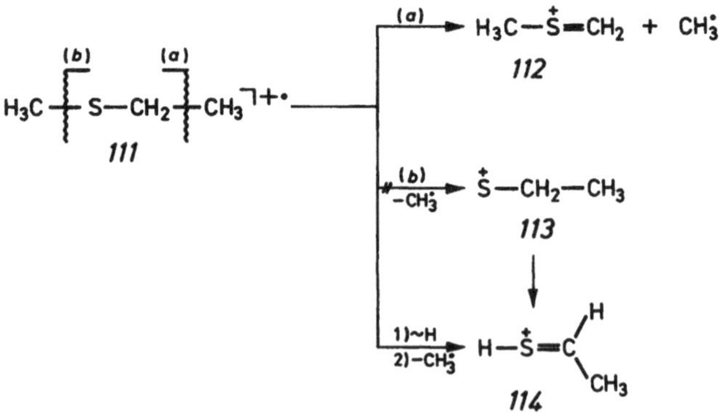

(25)

The question, whether *114* is formed directly from the molecular ion by an α-cleavage assisted by hydrogen migration or whether the ion is the result of a two-step process involving *113* as a transient species could be answered by investigating the ^{2}H-labelled sulfides *111a* and *111b*. From *111a* $CD_3^•$ and $CH_3^•$ are eliminated in the ratio 0.67:1. This value drops to 0.39:1 when the hydrogen atoms of the methylene group in *111a* are substituted by deuterium *111b* indicating that the cleavage of the C—S bond (type b in (25)) is subject to a primary isotope effect. However, a primary isotope effect is only operative when a hydrogen/deuterium migration is involved in the rate determining step of the cleavage. An explanation is given in (26) in which stretching of the (C—S) bond is supposed to be assisted by bridging through the adjacent hydrogen/deuterium (26). Whether *115* represents a transition state or a discrete intermediate is still open to question. Thus, from a mechanistic point of view it is still impossible to decide whether the (C—S) cleavage follows a one-step or a two-step mechanism.

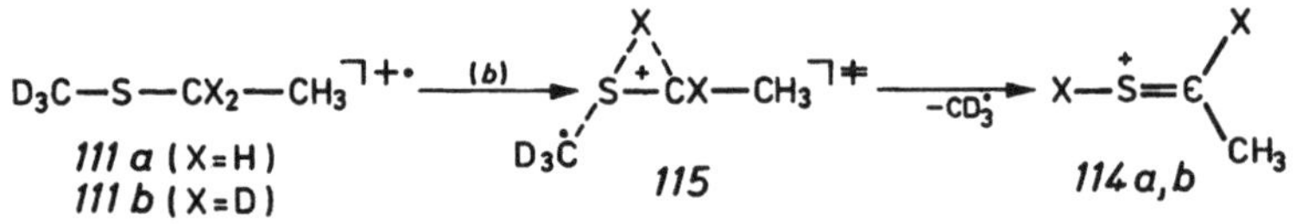

(26)

Similar results were obtained for other sulfides. For example, the investigation of labelled precursors or the analysis of CA spectra in combination with *ab initio*-calculations (carried out at the STO—3 G level) indicate that both isomeric CH_3S^+ ions *116* and *117* (the latter as a triplet) do exist in potential minima[52]. *116* is calculated to be 30 kcal · mol^{-1} more stable than the methyl-sulfenium ion *117*, and the barrier for the isomerization *117*→*116* is calculated to be as high as approximately 40 kcal · mol^{-1}. Ions *116* are formed *via* direct β-cleavage from the mercaptanes *118*, but the isomeric sulfenium ion *117* cannot be obtained by (S—R) cleavage of ionized *119*. In that case (R = alkyl) the dissociation is accompanied by hydrogen migration giving rise to the formation of the more stable sulfonium ion *116*.

However, *117* can be formed by ionization of those precursors in which the (S—R) bond is relatively weak, such as in disulfides *121* (27)[52].

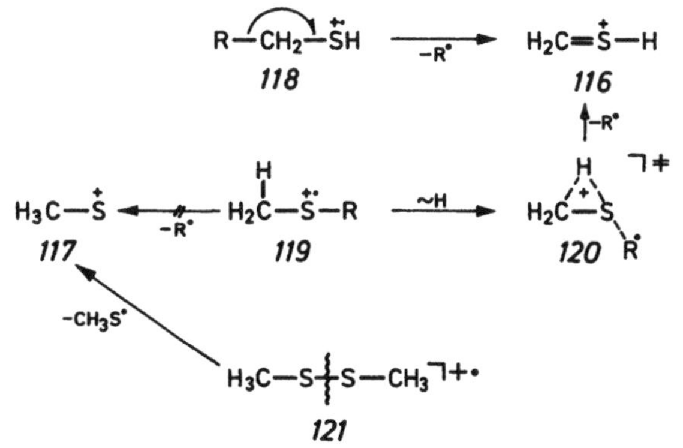

(27)

Similar results were obtained for the cleavage of the (S—C) bond in ionized *122*[51]. For *metastable* ion decompositions it is observed that the loss of $C_2D_5^{\cdot}$ is favoured over $C_2H_5^{\cdot}$ elimination by a factor of 2.8:1. This value reflects directly the operation of a primary kinetic isotope effect.

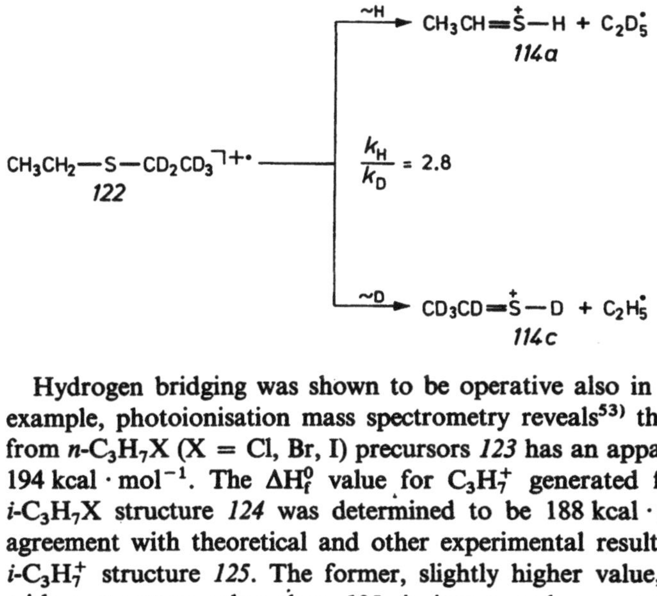

(28)

Hydrogen bridging was shown to be operative also in some other systems. For example, photoionisation mass spectrometry reveals[53] that the $C_3H_7^+$ ion formed from *n*-C_3H_7X (X = Cl, Br, I) precursors *123* has an apparent heat of formation of 194 kcal · mol^{-1}. The ΔH_f^0 value for $C_3H_7^+$ generated from precursors with an *i*-C_3H_7X structure *124* was determined to be 188 kcal · mol^{-1} which is in close agreement with theoretical and other experimental results and consistent with an *i*-$C_3H_7^+$ structure *125*. The former, slightly higher value, which is not consistent with a structure other than *125*, is interpreted as a result of a barrier for the cleavage of the C—X bond and the concomitant hydrogen migration *123*→*127*→*125*. There is no indication for the formation of the ion *126* (29).

A mechanism closely related to that of (29) can be viewed for the loss of methyl radicals from terminal positions of ionized *n*-alkanes[54]. For example, based on experiments with ^{2}H-labelled hydrocarbons and on careful analysis of energetic data it was shown convicingly that $CD_3^{\cdot}$ loss from *128* gives only the secondary cation *130*. From the fact that *131* is not found to be formed at all, it must be concluded that the cleavage of the (C—C) bond is associated with hydrogen bridging, *128*→*129*→*130* (30).

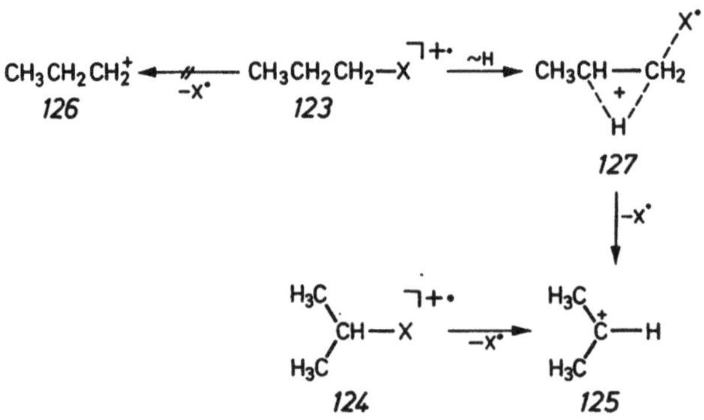

(29)

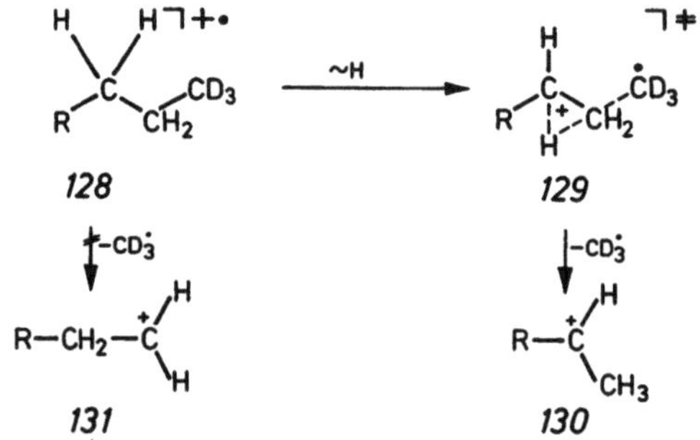

(30)

Convincing evidence for the operation of "hydrogen bridging" in the course of (C—C) bond dissociations was presented by *Howe*[55] who investigated the unimolecular loss of $CH_3^{\cdot}$ and $C_3H_7^{\cdot}$ from metastable molecular ions of 2-methylhexane *132*. ^2H-labelled analogues showed that the $CH_3^{\cdot}$ loss involves both (structurally equivalent) CH_3 groups attached to $C_{(2)}$. $C_3H_7^{\cdot}$ was found to be eliminated from the unbranched part of the molecule, involving the positions $C_{(4)}$, $C_{(5)}$, and $C_{(6)}$. The $CH_3^{\cdot}$ loss is very likely to occur as a simple (C—C) bond dissociation, thus giving rise to the formation of a secondary carbenium ion *133*, but the elimination of $C_3H_7^{\cdot}$ must follow a more complicated pathway despite the clean labelling results. This follows directly from simple energetic considerations. The fact that both dissociations are in competition (ratio of $C_3H_7^{\cdot}/CH_3^{\cdot}$ loss is 0.23) requires comparable activation energies. Assuming similar (or no) reverse activation energies the activation energy for the elimination of $C_3H_7^{\cdot}$ and formation of the primary carbocation *134* is approximately 23 kcal · mol higher than that for the $CH_3^{\cdot}$ loss *123*→*133* (31). This difference is much too high to observe *132*→*134* and *132*→*133* as competing reactions (31). However, if $C_3H_7^{\cdot}$ elimination is associated with a concomitant hydrogen migration, thus giving rise to the formation of the t-butyl cation

135 the energy gap is reduced to $\leqq 9$ kcal · mol^{-1} which makes a competition with the $CH_3^{\cdot}$ elimination more likely.

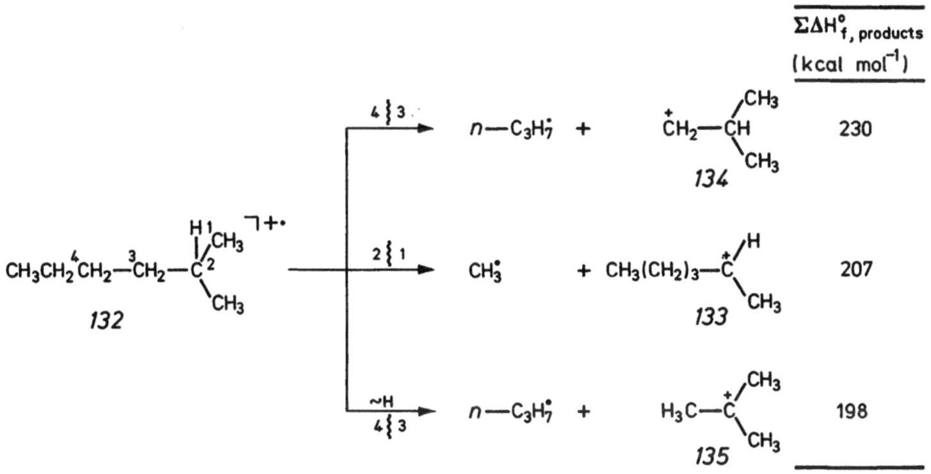

(31)

A nice, experimental confirmation that elimination of $C_3H_7^{\cdot}$ does not follow a simple (C—C) bond cleavage mechanism, but that is actually associated with hydrogen migration is provided by the investigation of the specifically ^{2}H-labelled hydrocarbon *132a*. The ratio of $C_3H_7^{\cdot}/CH_3^{\cdot}$ loss is 0.23 for the undeuterated compound, but the ratio drops to a value of 0.05 in the case of *132a*. Again, this decrease of $C_3H_7^{\cdot}$ elimination is a direct consequence of a primary kinetic isotope effect operating in the transition state of the rearrangement/dissociation process *132a→136→135a* (32).

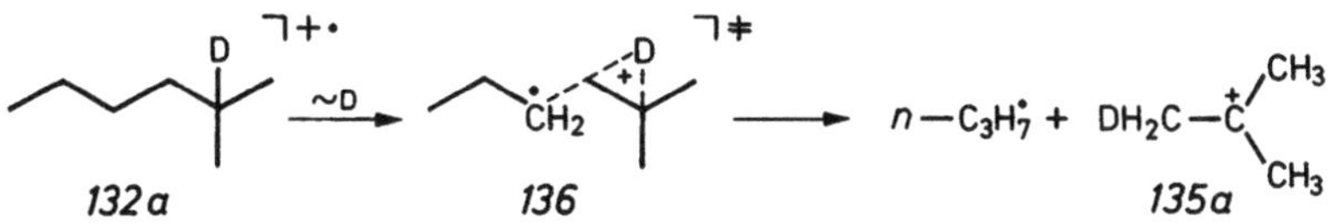

(32)

An unusual example of an anchimerically assisted (O—C) bond cleavage is found in the electron impact induced dissociation of alkyl silylmethyl ether *137* (33)[56].

The experimental observations that 1) the relative abundances of the $[M-R^3]^+$ ions *increase* substantially with *decreasing* ionizing energy (see (33) for $R^1 - R^2$ = biphenylylen) and that 2) abundant metastable peaks are observed for R^3 loss from *137* are not consistent with a simple bond cleavage producing an oxenium-ion *138* (34). The results indicate that a rearrangement is involved in the dissociation step. We propose that the $Si(CH_3)_3$ group migrates to the oxygen atom *137→139*, and that it is the coordination of the empty Si orbitals with the electron rich oxygen atom which can be expected to lower the transition state energy.

That the silylated oxonium ion *140* is indeed the product formed upon cleavage of the (O—C) bons in *137* is demonstrated by CA spectroscopy. The CA spectra of the $[M—R^3]^+$ ions ($R^3 = CH_2Ph$) derived from *137* and generated independently from *141* via α-cleavage are identical and not sensitive to the energy of the ionizing electrons. The decisive importance of the coordination step *137*→*139* in lowering the transition state energy *via* (Si—O) bond formation can be seen when the $Si(CH_3)_3$ group is replaced by a group which is sterically comparable but electronically quite different. Such a functional group is, for instance, t-butyl. A comparative study was carried out for the three precursors *142* (35). Despite the fact that the radical R^3

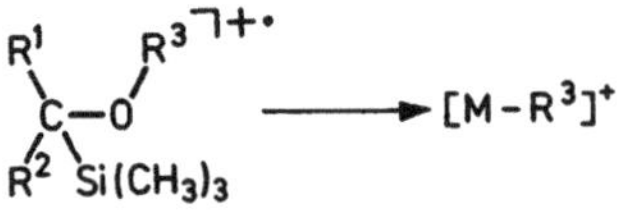

137 (R^1,R^2: aryl, alkyl)

R^3	% Σ_{40} for $[M-R^3]^+$ 70 eV	12 eV
$PhCH_2$	31	94
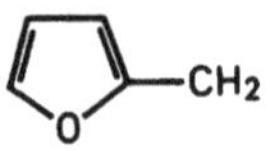	22	97
$CH_2{=}CH{-}CH_2$	17	94
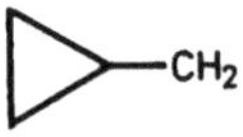	28	82
$Si(CH_3)_3CH_2$	20	90
CH_3	23	82

(33)

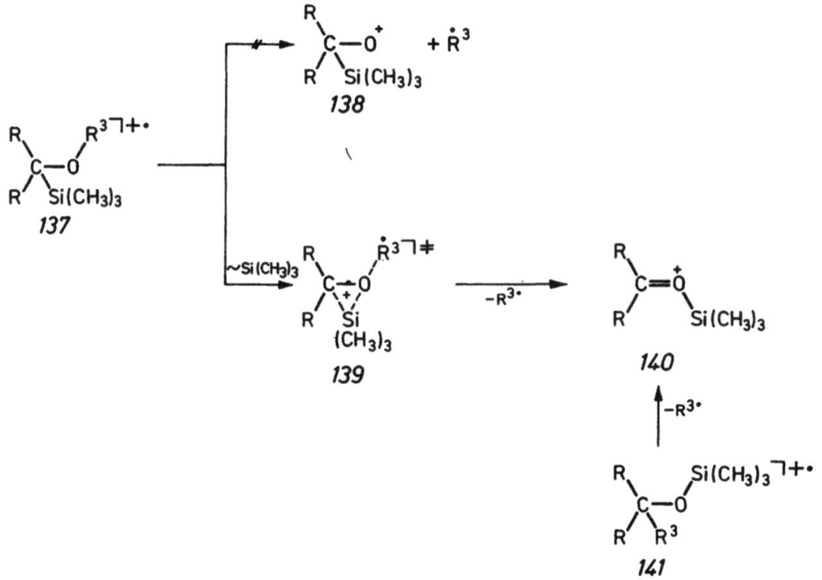

(34)

to be eliminated is quite stable ($R^3 = CH_2Ph$), the relative abundance of the $[M-CH_2Ph]^+$ ion is negligible (<0.1%) for X = $C(CH_3)_3$. This is a direct consequence of the absence of any anchimeric assistance in the transition state *143* in the case of the t-butyl derivative. Abundant $[M-PhCH_2]^+$ ions are recorded for substituents being able to coordinate, such as Si and Ge containing groups (35).

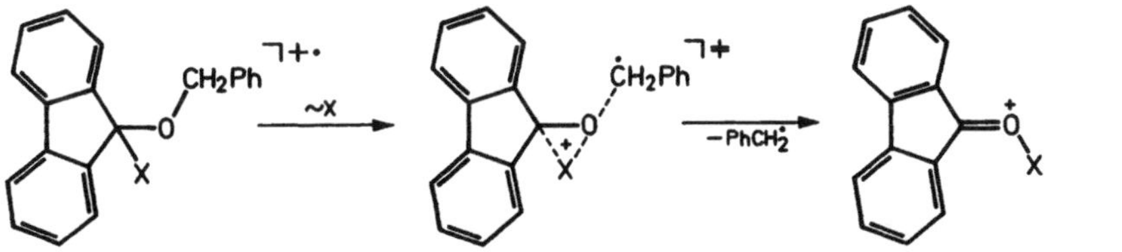

(35)

142 *143* *144*

X	% Σ_{40} for ion *144* 70 eV	12 eV
$C(CH_3)_3$	< 0.1	< 0.1
$Si(CH_3)_3$	31	94
$Ge(CH_3)_3$	15	84

6 Miscellaneous

6.1 Halogen Loss from Cation Radicals of Substituted Cyclopropanes[57)]

Electron impact ionization of both disubstituted cyclopropanes *145* and *146* lead to the elimination of the Br˙ radical. From the CA spectra of the resulting $[M-Br]^+$ ions and those of $C_4H_7^+$ ions generated from other suitable precursors it must be concluded that both cyclopropane derivatives give only the 1-methylallyl cation *147*. There is no experimental evidence for the generation of the isomeric 2-methylallyl cation *148*. Moreover, it was shown that *147* and *148* do not interconvert under the experimental conditions used (36).

At first sight the formation of *147* from *145* might be explained by a mechanism in which stretching of the (C—Br) bond is coupled with a symmetry allowed

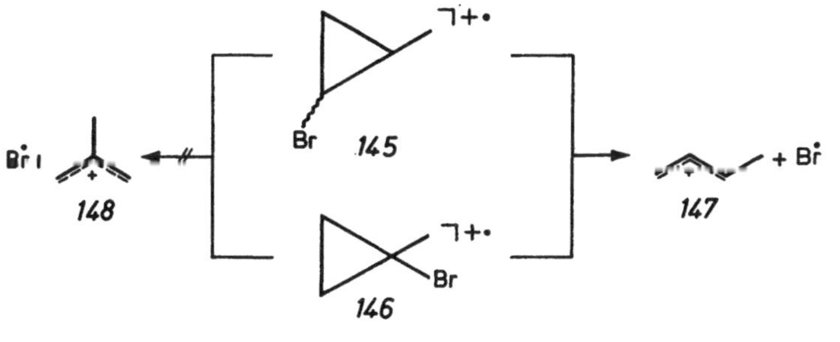

148 *145* *146* *147*

(36)

disrotatory ring opening of the incipient 2-methylcyclopropyl cation *149* to give the $C_4H_7^+$ product *147* (37). However, this mechanism is of no (or minor) importance for the dissociation of *145* as will be shown later.

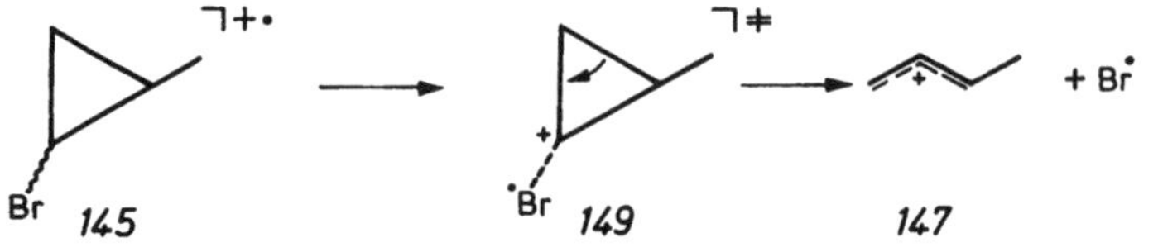

(37)

A similar mechanism for Br˙ loss from *146* would result in the formation of the 2-methylallyl cation *148*, but not of the required structure of the 1-methylallyl cation *147*. Formation of *147* from the intermediate *150* may be conceivable only if the activation energy for the ring opening of the incipient 1-methylcyclopropyl cation to give the 2-methylallyl cation *150*→*148* is higher than the energy required for a combined process of a [1.2]-hydride shift *150*→*151* and a subsequent symmetry allowed isomerization to *147*. However, quantum mechanical calculations[57] carried out at both a semi-empirical (MNDO) and an *ab initio* level (STO-3G/4-31G) clearly indicate that such an order of activation energies is not met (38). Thus, the experimentally established formation of a 1-methylallyl cation *147*, must proceed via a reaction pathway which is fundamentally different from all the mechanistic alternatives discussed so far in this section. It will be shown, that according to MNDO calculations a "hidden" hydrogen transfer reaction plays a decisive role in the halogen loss from ionized cyclopropane derivatives.

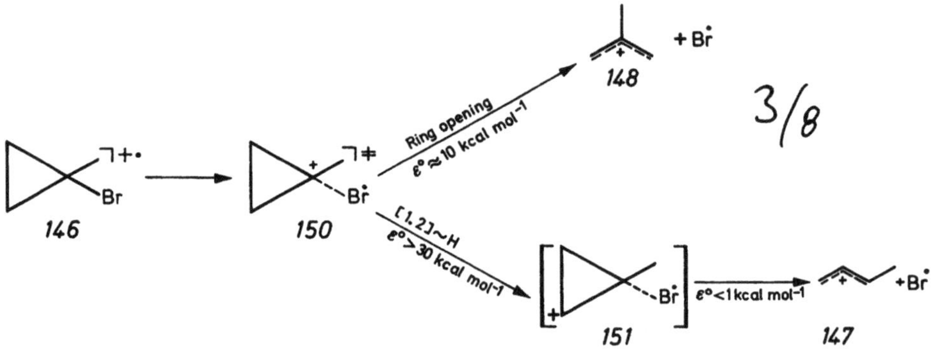

(38)

We have employed Dewar's semi-empirical MNDO-procedure[58] to compute the potential energy surface for halogen loss from ionized *chloro* substituted methylcyclopropanes *152* and *153*[59]. The very extensive calculations reveal that 1) direct halogen loss according to *145*→*149* or *146*→*150* (Cl˙ substituted for Br˙) is not favoured energetically and that 2) the cation radicals of substituted cyclopropanes in general do not exist in potential minima. They undergo spontaneous ring opening, thus giving rise to the formation of ionized acyclic isomers. For *152* is was found that the reaction sequence *152*→*154*→*155*→*147* (39) is energetically the most feasible pathway. This multistep process which involves ring opening, hydrogen migration and cleavage of an allylically activated (C—C) bond requires 21 kcal · mol^{-1} *less* energy than the alternative reaction *152*→*156*→*148* which commences with cleavage of the ($C_{(2)}$—$C_{(3)}$) bond of *152*.

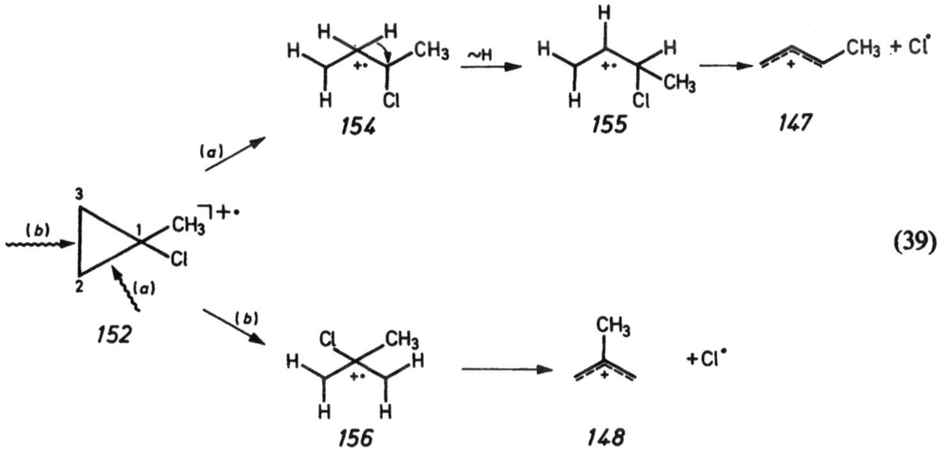

(39)

Part of the (two-dimensional) potential energy diagram for the isomerization and decomposition reactions of ionized *152* is shown in Fig. 1. Obviously, the multistep pathway *152*→→→*147* (39) involving the "hidden" hydrogen migration is by far the energetically most favoured reaction channel. In particular, it is worth mentioning that direct Cl˙ elimination for the ringopened intermediate *154* is extremely unlikely. The calculations reveal that continuous stretching of the (C—Cl) bond in *154* proceeds *via* a transition state, which formally corresponds with a cationic carbene which is weakly connected with the chlorine radical. The heat of formation of this complex

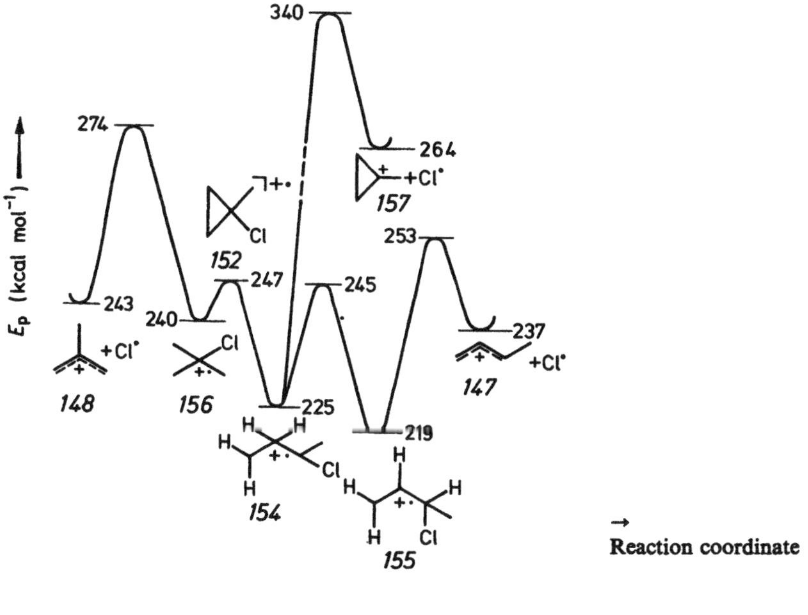

is calculated to be 340 kcal · mol^{-1} and further progress on the reaction coordinate results in the formation of the 1-methylcyclopropyl cation *157*.

Similar computational results are obtained for the cation radicals of the 1.2-disubstituted cyclopropane *153*. The calculations indicate that two energetically comparable pathways can account for the formation of the 1-methylallyl cation *147*. One involves a multistep reaction commencing with spontaneous cleavage of the ($C_{(1)}$—$C_{(2)}$) bond of *153*. This is followed by the hydrogen migration *158*→*159* and terminated by Cl˙ loss. The energy for the highest point of the reaction coordinate is calculated to be 255 kcal · mol^{-1}. This transition state energy is only 2 kcal · mol^{-1} higher than that for direct Cl˙ loss from the intermediate *160* which is obtained after cleavage of the ($C_{(2)}$—$C_{(3)}$) bond of *153*. The alternative reaction sequence *153*→*161*→*162* →*148* (40) is calculated to require an activation energy approximately 12–14 kcal · mol^{-1} higher than those for the processes discussed above. This difference may be sufficient to suppress the formation of the 2-methylallyl cation *148*, which is in agreement with the experimental observations.

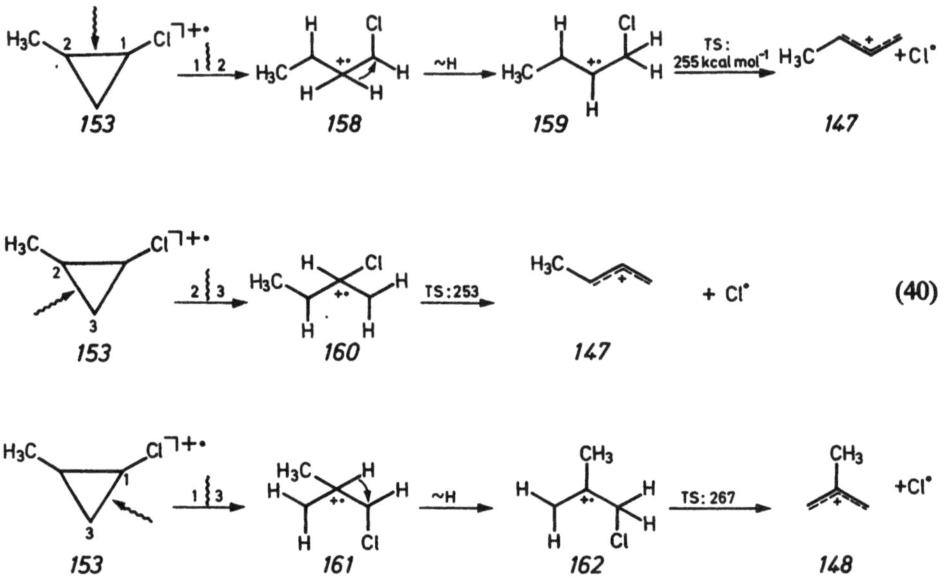

(40)

6.2 Hydroxyl Loss from the Ionized Enol of Acetic Acid

It is well known that neutral simple ketones *163* in general are *more stable* than their isomeric enol form *164*[60,61]. However, for the corresponding cation radicals *165* and *166* it has been shown both experimentally[62,63] and computationally[64,65] that, depending on the substituent R, the ionized ketones *165* are significantly *less stable* than their isomeric enol cation radicals *166* (41). Moreover, CA spectroscopy reveals that both species exist in potential minima and ICR spectroscopy demonstrate quite different reactivities of the ions *165* and *166*[66].

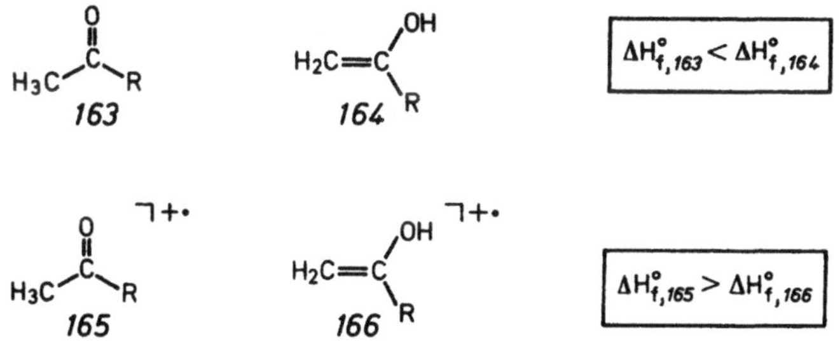

(41)

Particularly interesting is the mechanism for the unimolecular hydroxyl loss from ionized acetic acid *167* and from its enol form *168*[63,67,68]. ^{2}H-labelling and energetic data reveal that OH$^{\cdot}$ loss from *167* can be described by an α-cleavage process, producing the acylium ion *169*. However, OH$^{\cdot}$ loss from *168* follows a fundamentally different mechanism. Both the energetics of the latter reaction and CA spectroscopic data and the labelling data are not found to be consistent with the formation of an α-cleavage product, that is an O-protonated ketene *170*. All data indicate that the product generated by OH$^{\cdot}$ loss from *168* also has the structure of the acylium ion *169* (42).

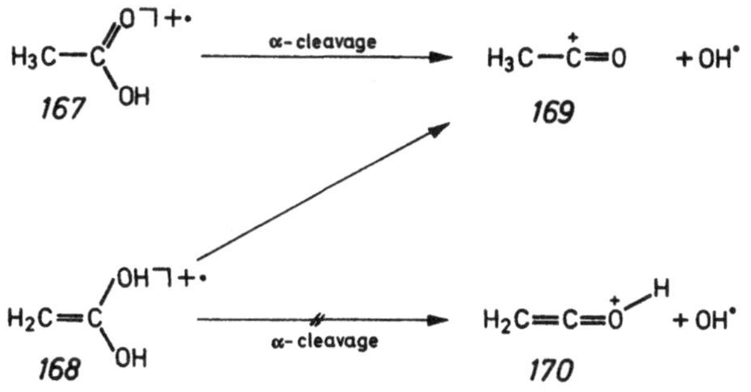

(42)

The investigation of the ^{2}H-labelled enol *168a* clearly shows the existence of a primary isotope effect in the loss of hydroxyl. The ratio for loss of OD$^{\cdot}$/OH$^{\cdot}$ is determined experimentally to be 71:29. This result is best compatible with a two-step mechanism for hydroxyl loss from ionized enol *168* (43). Isomerization of the enol *168* to ionized acetic acid *167* occurs in the rate-determining step and it is this rearrangement in which the isotope effect is operative (retardation of deuterium transfer compared with hydrogen migration, thus discriminating against OH$^{\cdot}$ elimination in the final dissociation step).

The above mentioned conclusion is confirmed nicely by energetic measurements[63] (Fig. 2). The experimentally derived transition state energy for OH$^{\cdot}$ loss from *168* is approximately 175 kcal · mol^{-1}. This is much too low for a direct α-cleavage process which would give rise to the formation of *170* and OH$^{\cdot}$, the sum of the heats of formation of which is around 190 kcal · mol^{-1}. However, the transition state energy for the isomerization process *168*→*167* is higher than the energy for the dissociation step, *167*→*169* (175 compared with 160 kcal · mol^{-1}). Consequently, the difference between these transition state energies will be available as non-fixed energy, $\varepsilon^{\neq}$, when the ions pass over the transition state for dissociation. The part of $\varepsilon^{\neq}$ which will flow into the reaction coordinate as translational energy should cause line broadening of the metastable peak. This is observed indeed: the average kinetic energy release associated with OH$^{\cdot}$ loss from *168* (*via 167*) is significantly larger (T = 2.2 kcal · mol^{-1}) than that for direct OH$^{\cdot}$ elimination from ionized acetic acid *167* (T = 0.3 kcal · mol^{-1}).

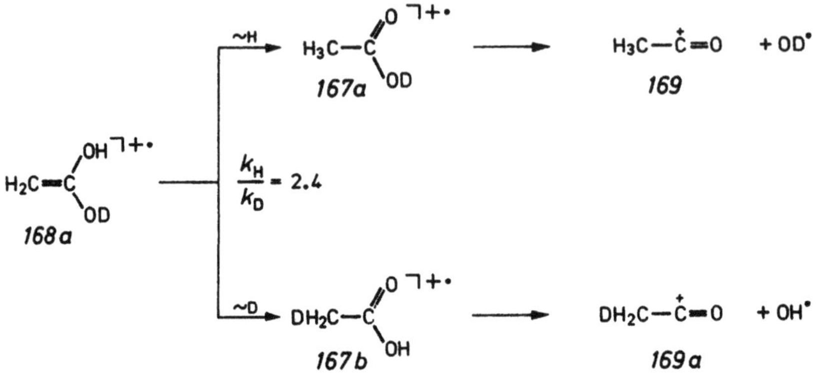

(43)

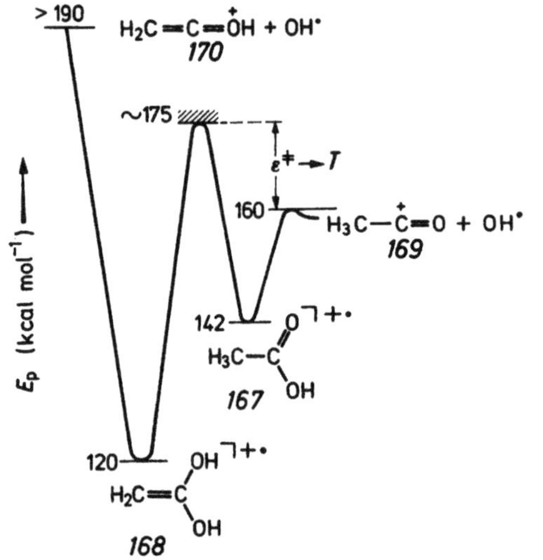

6.3 Benzyl Radical Elimination from Difunctionalized Alkane Cation Radicals

Hesse[69] has shown that the unimolecular chemistry of cation radicals of di- or polyfunctionalized alkanes is strongly dependent upon the interaction of the functional groups, mainly *via* neighbouring group participation in the transition states and the formation of cyclic product ions. However, two nice examples of such interactions in which "hidden" hydrogen migrations are of special importance were reported by Nibbering[70,71]. The benzyl radical elimination from ionized α,ω-benzyloxy benzylamino alkanes *171* was shown not to be the result of the direct (O—C) bond cleavage *171*→*172*, but of the two step process *171*→*173*→*174* (44). The reason that the two-step process is favoured in comparison with the direct cleavage reaction is exactly the same as discussed in Chapter IV.

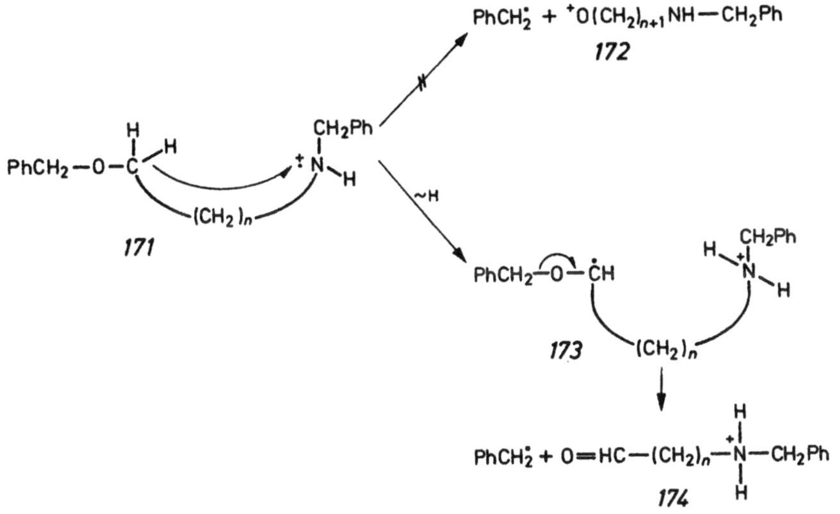

(44)

Similarly it has been observed[71] that the loss of the benzyl radical from the molecular ions of 1.4-dibenzyloxybutane *175* is initiated by successive migration of a benzylic hydrogen atom to the opposite ether function *175*→*176* and transfer of the benzyl cation from this protonated oxygen atom to the uncharged oxygen atom in an S_Ni-type reaction *176*→*177*. It is the intermediate *177* from which $PhCH_2^{\bullet}$ is eventually eliminated in a double-bond formation process giving rise to the formation of the oxonium ion *178* (45). Possibly many other dissociation processes of polyfunctionalized alkane cation radicals belong to the same category.

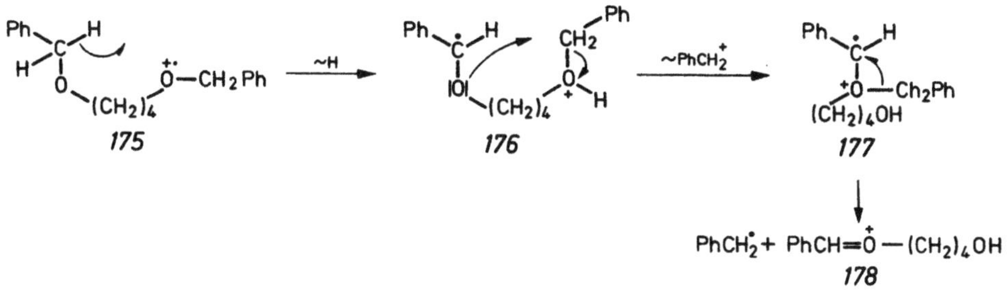

(45)

6.4 Hydroxyl Loss from Cation Radicals of Substituted Aldoximes

OH˙ elimination from *ortho* substituted aldoximes *179* (X = CH_2, NH, O) may be at least partially the result of a hydrogen migration/cyclization/elimination process, whereby the heterocycles *182* are formed[72] (46). A metastable peak shape analysis, the investigation of ^{2}H-labelled derivatives and the study of positional isomers indicate that in addition to *182* the protonated isocyanide *183* is formed *via* a mechanism which is not fully understood. However, it is known that the generation of *183* occurs without any detectable interaction with the XH *ortho* substituent.

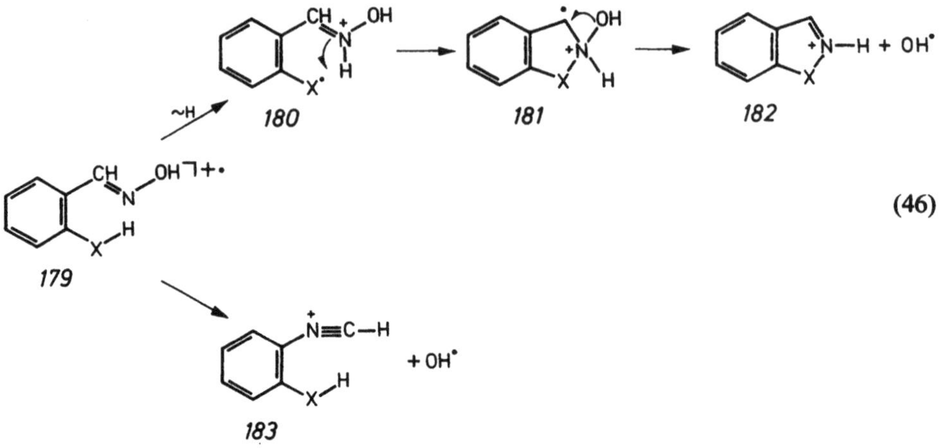

(46)

6.5 Hydrogen Radical Elimination from Ionized Ethylene

Last but not least it should be mentioned that the process of H˙ loss from ionized ethylene *184* is a reaction which not only follows a simple (C—H) dissociation giving rise to the formation of the classical vinyl cation *31* (47). From the careful analysis of extensive *ab initio* calculations it has to be concluded[73] that the stretching and dissociation of the (C—H) bond of ionized *184* is also coupled with a molecular

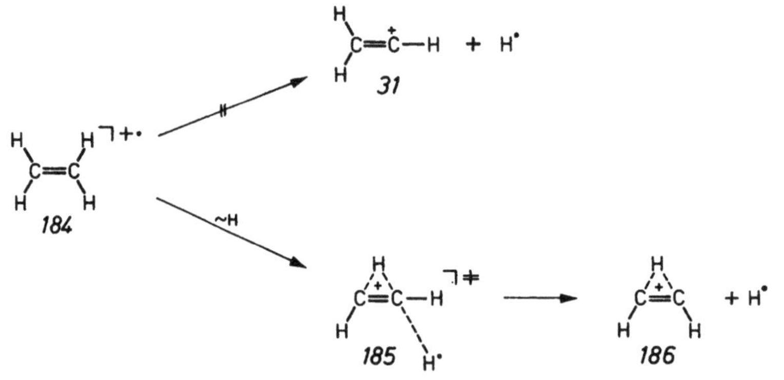

(47)

motion in which the hydrogen bridging occurs, *184*→*185*. Despite the fact that *185* cannot be viewed as a structurally well-defined species existing in a potential minimum, this example and also those discussed above indicate that many radical eliminations from gaseous cation radicals are either preceded by "hidden" hydrogen migrations (two- or multistep-reactions) or anchimerically assisted processes (one-step reactions) in which hydrogen atoms play a decisive role as bridging elements.

7 Acknowledgements

The support of our own work (cited in the references) by the Fonds der Chemischen Industrie, Frankfurt, the Deutsche Forschungsgemeinschaft, Bonn (Projects Schw 221/2, 4, 5) and the Gesellschaft von Freunden der Technischen Universität Berlin is gratefully acknowledged. I am very pleased to thank my coworkers (mentioned in the references) for both their experimental and conceptual contributions. The exchange of valuable information with Prof. *A. Mandelbaum*, Haifa, and the assistance of Prof. *N. M. M. Nibbering*, Amsterdam, in preparing the English version of the manuscript are kindly acknowledged.

8 References

1. Kingston, D. G. I., Hobrock, B. W., Bursey, M. M., Bursey, J. T.: Chem. Rev. *75*, 693 (1975)
2. Schwarz, H.: Top. Curr. Chem. *73*, 232 (1978)
3. Kingston, D. G. I., Bursey, J. T., Bursey, M. M.: Chem. Rev. *74*, 215 (1974)
4. Gillis, R. G., Occolowitz, J. L., Pirani, J. F.: Org. Mass Spectrom. *2*, 425 (1969)
5. Bohlmann, F., Herrmann, R., Schwarz, H., Schiebel, H. M., Schröder, N.: Tetrahedron *33*, 357 (1977)
6. Schwarz, H., Sezi, R., Rapp, U., Kaufmann, H., Meier, S.: Org. Mass Spectrom. *12*, 39 (1977)
7. For an earlier review concerning cyclization type reactions see: Cooks, R. G.: Org. Mass Spectrom *2*, 481 (1969)
8. Schaldach, B., Grützmacher, H. F.: Int. J. Mass Spectrom, Ion Phys. *31*, 271 (1979), and references cited therein
9. Köppel, C., Van de Sande, C. C., Nibbering, N. M. M., Nishishita, T., McLafferty, F. W.: J. Am. Chem. Soc. *99*, 2883 (1977)
10. Shapiro, R. M., Tomer, K. B.: Org. Mass Spectrom *3*, 333 (1970)
11. For a preliminary report on "hidden" hydrogen rearrangements see: Schwarz, H.: Nachr. Chem. Techn. Lab. *28*, 158 (1980); Bar-Shai, R., Bortinger, A., Sharvit J., Mandelbaum, A.: Isr. J. Chem. *20*, 137 (1980).
12. Wesdemiotis, C., Schwarz, H.: Angew. Chem. *90*, 724 (1978); Angew. Chem. Int. Ed. Engl. *17*, 678 (1978)
13. Weisz, A., Mandelbaum, A.: J. Chem. Soc. Chem. Commun. *1978*, 521
14. Hemberger, P. H., Kleingeld, J. C., Levsen, K., Mainzer, N., Mandelbaum, A., Nibbering, N. M. M., Schwarz, H., Weber, R., Weisz, A., Wesdemiotis, C.: J. Am. Chem. Soc., *102* 3736 (1980)
15. Beckey, H. D.: Principles of Field Ionization and Field Desorption Mass Spectrometry. Oxford: Pergamon Press 1977
16. Williams, D. H., Howe, I.: Principles of Organic Mass Spectrometry. London: McGraw Hill 1972
17. Levsen, K.: Fundamental Aspects of Organic Mass Spectrometry. Weinheim: Verlag Chemie 1978

18. Levsen, K., Schwarz, H.: Angew. Chem. *88*, 589 (1976); Angew. Chem. Int. Ed. Engl. 15, 509 (1976)
19. Richter, W. J., Schwarz, H.: Angew. Chem. *90*, 449 (1978); Angew. Chem. Int. Ed. Engl. *17*, 424 (1978)
20. McAdoo, D. C., Witiak, D. N., McLafferty, F. W., Dill, J. D.: J. Am. Chem. Soc. *100*, 6639 (1978)
21. McLafferty, F. W. in McLafferty, F. W. (Ed.): Mass Spectrometry of Organic Ions. New York: Academic Press 1963
22. Meyerson, S.: Int. J. Mass Spectrom. Ion Phys. *1*, 309 (1968)
23. Van de Sande, C. C., DeMeyer, C., Maquestiau, A.: Bull. Soc. Chim. Belge *85*, 79 (1976)
24. Cason, J., Khodair, A. I. A.: J. Org. Chem. *31*, 3618 (1966)
25. Mandelbaum, A. in Kagan, H. (Ed.): Stereochemistry. Stuttgart: Georg Thieme Verlag 1977, p. 165
26. Herrmann, R., Schwarz, H.: Z. Naturforsch. *31b*, 1013 (1976)
27. Burgers, P. C., Terlouw, J. K., Vijfhuizen, P. C., Holmes, J. L.: Org. Mass Spectrom. *13*, 470 (1978)
28. Terlouw, J. K., Burgers, P. C., Schwarz, H.: Org. Mass Spectrom *15*, 599 (1980)
29. Reetz, M. T., Maier, W. F.: Theor. Chim. Acta. *35*, 163 (1974)
30. Bursey, M. M., Harvan, D. J., Parker, C. E., Pedersen, L. G., Hass, J. R.: J. Am. Chem. Soc. *101*, 5489 (1979)
31. Maier, W. F.: Personal communications, March 1980
32. Seibl, J.: Massenspektrometrie. Frankfurt/Main: Akademische Verlagsanstalt 1970
33. Schwarz, H., Bornowski, H., Petruck, G. M.: Org. Mass Spectrom. *10*, 469 (1975)
34. Naito, T.: Tetrahedron *24*, 6237 (1968)
35. Budzikiewicz, H., Djerassi, C., Jackson, A. H., Kenner, G. W., Newman, D. J., Wilson, J. M.: J. Chem. Soc. *1964*, 1949
36. Grigg, R., Sargent, M. V., Williams, D. H., Knight, J. A.: Tetrahedron *21*, 3441 (1965)
37. Bursey, M. M., Hass, J. R., Harvan, D. J., Parker, C. E.: J. Am. Chem. Soc. *101*, 5485 (1979)
38. Schwarz, H., Wolfschütz, R.: Org. Mass Spectrom. *11*, 773 (1976)
39. Schwarz, H.: Unpublished results
40. Herrmann, R., Schwarz, H.: Z. Naturforsch. *31b*, 870, 1667 (1976)
41. Barlow, M. G., Haszeldine, R. N., Kershaw, M. J.: J. Chem. Soc. Perkin I, *1974*, 1736; *1975*, 2005
42. Cram, D. J., Reeves, R. H.: J. Am. Chem. Soc. *80*, 3094 (1958)
43. Vögtle, R., Neumann, P.: Topics Curr. Chem. *48*, 67 (1974)
44. Grützmacher, H. F.: personal communication, April 1980
45. Beynon, J. H., Job, B. E., Williams, A. E.: Z. Naturforsch. *20a*, 180 (1965)
46. Meyerson, S., Corbin, J. L.: J. Am. Chem. Soc. *87*, 3045 (1965)
47. Neeter, R., Nibbering, N. M. M.: Org. Mass Spectrom. *5*, 735 (1971)
48. Neeter, R., Nibbering, N. M. M.: Tetrahedron *28*, 2575 (1972)
49. Van de Graaf, B., McLafferty, F. W.: J. Am. Chem. Soc. *99*, 6806 (1977)
50. Broer, W. J., Weringa, W. D., Nieuwpoort, W. C.: Org. Mass Spectrom. *14*, 543 (1979)
51. However, recent *ab initio* calculations (carried out at the STO-3G level) indicate that *113* does exist in a shallow minimum. The rearrangement to *114* affords a few kcal/mol: Van de Graaf, B.: Personal communication, April 1980; see also: Van de Graaf, B., Potters, J. J. M., Schuyl, P. J. W.: Advanc. Mass Spectrom. *8a*, 678 (1980)
52. Dill, J. D., McLafferty, F. W.: J. Am. Chem. Soc. *101*, 6526 (1979)
53. Traeger, J. C.: Int. J. Mass Spectrom. Ion Phys. *32*, 309 (1980)
54. Wolkoff, P., Holmes, J. L.: J. Am. Chem. Soc. *100*, 7346 (1978)
55. Howe, I.: Org. Mass Spectrom. *10*, 767 (1975)
56. Schwarz, H., Wesdemiotis, C., Reetz, M. T.: J. Organomet. Chem. *161*, 153 (1978)
57. Bowen, R. D., Chandrasekhar, J., Frenking, G., Schleyer, P. v. R., Schwarz, H., Wesdemiotis, C., Williams, D. H.: Chem. Ber. *113*, 1084 (1980)
58. Dewar, M. J. S., Thiel, W.: J. Am. Chem. Soc. *99*, 4899, 4907 (1977)
59. For the corresponding bromo derivatives at present no parameters are available to carry out the calculation

60. Forsén, S., Nilsson, M. in Patai, S. (Ed.): The Chemistry of the Carbonyl Group, II. New York: Wiley-Interscience 1970, p. 157
61. Hart, H.: Chem. Rev. *79*, 515 (1979)
62. Holmes, J. L., Terlouw, J. K., Lossing, F. P.: J. Phys. Chem. *80*, 2860 (1976)
63. Holmes, J. L., Lossing, F. P.: J. Am. Chem. Soc., *102*, 1591 (1980)
64. Bouma, W. J., MacLeod, J. K., Radom, L.: J. Am. Chem. Soc. *101*, 5540 (1979)
65. Apeloig, Y.: private communication, April 1979
66. Schwarz, H.: Nachr. Chem. Techn. Lab. *26*, 792 (1978)
67. Levsen, K., Schwarz, H.: J. Chem. Soc. Perkin Trans, *2*, *1976*, 1231
68. Schwarz, H., Williams, D. H., Wesdemiotis, C.: J. Am. Chem. Soc. *100*, 7052 (1978)
69. Bosshardt, H., Hesse, M.: Angew. Chem. *86*, 256 (1974); Angew. Chem. Int. Ed. Engl. *13*, 252 (1974)
70. Bruins, A. P., Nibbering, N. M. M.: Org. Mass Spectrom. *11*, 271 (1976)
71. Bruins, A. P., Nibbering, N. M. M.: Tetrahedron *30*, 493 (1974)
72. Vijfhuizen, P. C., Terlouw, J. K.: Org. Mass Spectrom. *11*, 888 (1976); *12*, 63, 245 (1977)
73. Lorquet, J. L.: Plenary lecture presented at the EUCHEM conference on "The Chemistry of Ion Beams", Lunteren, April 1980. See also: a) Lorquet, J. C., Sannen, C., Raşeev, G. J. Am. Chem. Soc. *102*, 7936 (1980); b) Sannen, C., Raşeev, G., Galloy, C., Fauville, G., Lorquet, J. C., J. Chem. Phys. *74*, 2402 (1981)

Recent Advances in Thiepin Chemistry

Ichiro Murata and Kazuhiro Nakasuji

Department of Chemistry, Faculty of Science, Osaka University,
Toyonaka, Osaka 560, Japan

Table of Contents

1 Introduction

Seven-membered conjugated systems having a hetero atom, heteropin (*1*), consist of 1*H*-azepine (*1*, X = NH), oxepin (*1*, X = O), and thiepin (*1*, X = S). Since these divalent hetero atoms are isoelectronic with the ethylenic linkage, heteropins might be considered to be 8π electron heteroannulenes which are not aromatic but antiaro-

1

matic according to the Hückel rule. In this context, the chemistry of the heteropins has long been a subject of great interest to both synthetic and theoretical chemists.

In connection with the chemistry of the reactive transient species, nitrene, the chemistry of azepines is well documented [1]. Also, the chemistry of oxepins has been widely developed due to the recent interest in the valence isomerization between benzene oxide and oxepin and in the metabolism of aromatic hydrocarbons [2]. In sharp contrast to these two heteropins, the chemistry of thiepins still remains an unexplored field because of the pronounced thermal instability of the thiepin ring due to ready sulfur extrusion. Although several thiepin derivatives annelated with aromatic ring(s) have been synthesized, the parent thiepin has never been characterized even as a transient species [3].

At present, the problems in thiepin chemistry awaiting solution are (i) how to construct the thiepin skeleton under mild reaction conditions, (ii) what are the structural effects on thermal stability of thiepin, (iii) whether the thianorcaradiene is an intermediate of sulfur extrusion reaction or not, (iv) what is the molecular structure of the thiepin (planar or nonplanar), (v) what is the antiaromaticity of the thiepin ring.

The aim of this article is to present a survey of the recent developements in the chemistry of thiepin mainly from our own studies. The principal information available prior to 1970 is already included in former reviews [4], and interested readers are recommended to refer to these articles. At first we would like to briefly summarize the previously available methods of synthesizing the thiepin skeleton.

2 Synthesis of Annelated Thiepins

2.1 By Elimination Reaction

This synthetic approach involves elimination reactions of hydrothiepins having the appropriate leaving group(s). Since fairly drastic conditions have to be employed, this method is restricted solely to the preparation of stable thiepin derivatives.

Traynelis et al. reported that when 2,3-dihydrobenzo[*b*]thiepin (*2*) was refluxed in petroleum ether with sulfuryl chloride and sodium hydrogen carbonate, the products isolated and characterized were naphthalene and sulfur [5]. The origin of these products was rationalized by α-chlorination of *2* and loss of hydrogen chloride to form benzo[*b*]thiepin (*4*) as an intermediate which underwent sulfur extrusion. However, when *2* and an equivalent of N-chlorosuccinimide were allowed to react at room

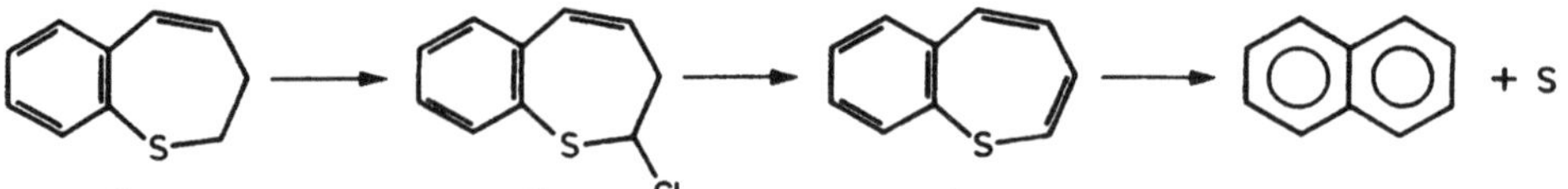

2 3 4

temperature, the α-chlorinated product *3* was obtained which, on treatment with potassium *tert*-butoxide in dimethyl sulfoxide provided benzo[*b*]thiepin (*4*) as a yellow oil [6]. The same treatment of 2,2-dichloro-2,3-dihydrobenzo[*b*]thiepin (*5*), obtained from *2* with two equivalents of N-chlorosuccinimide, with potassium *tert*-butoxide gave 2-chlorobenzo[*b*]thiepin (*6*) [6]. However, the elimination reaction under more

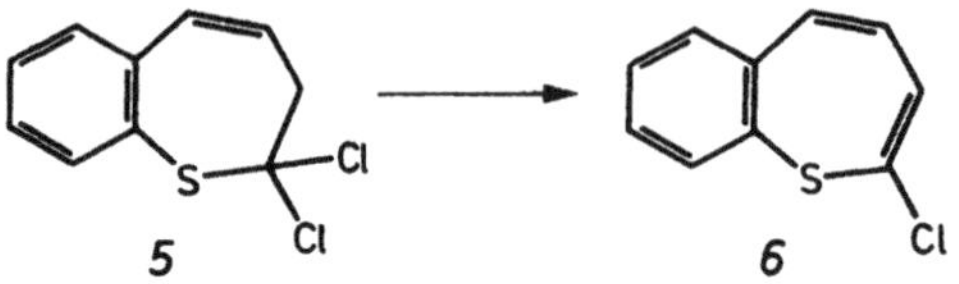

5 6

drastic conditions, such as the pyrolysis of 5-acetoxy-2-chloro-4,5-dihydrobenzo[*b*]-thiepin (*7*), gave only 1-chloronaphthalene [5] instead of *6*.

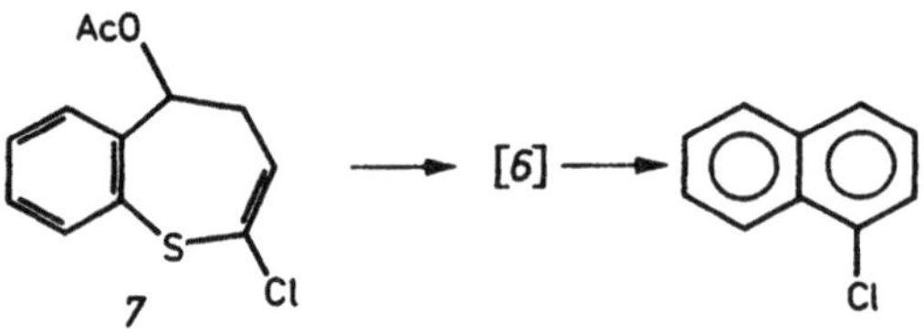

7

Thiepins stabilized by two annelated benzene rings could generally be prepared by the usual elimination processes, and some examples are shown in the following scheme [7,8,9,10].

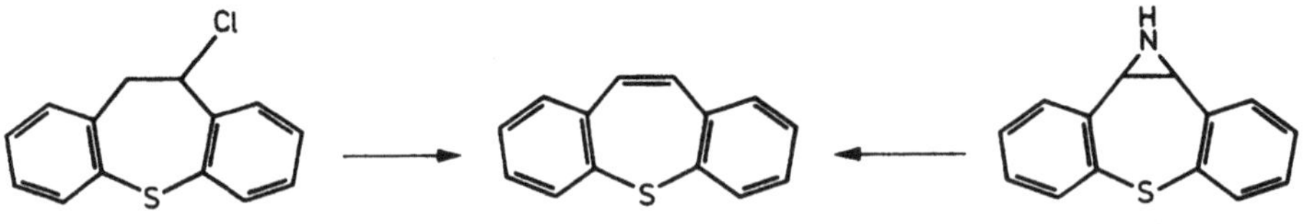

8 9 10

The Pummerer reaction has successfully been applied to the synthesis of some novel thiepins (*13*) starting from the 4,5-dihydro derivatives *11* and proceeding via the sulfoxide *12* [11,12].

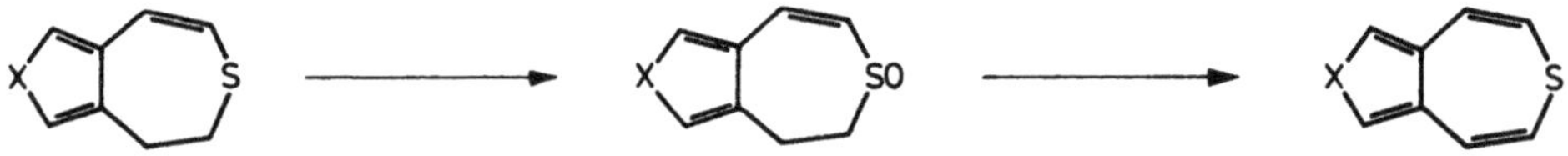

11 12 13

2.2 By Enol Fixation

In order to construct the thiepin conjugation, the seven-membered cyclic ketones containing sulfur could be converted to either their enol ethers or enol acetates. The resulting thiepins should have a number of substituents. The first stable 3,5-diacetoxy-4-phenylbenzo[*b*]thiepin (*15*) has been obtained by Hofmann et al. from the diketone (*14*) by acetylation with acetic anhydride in pyridine in good yield [13]. By this methodo-

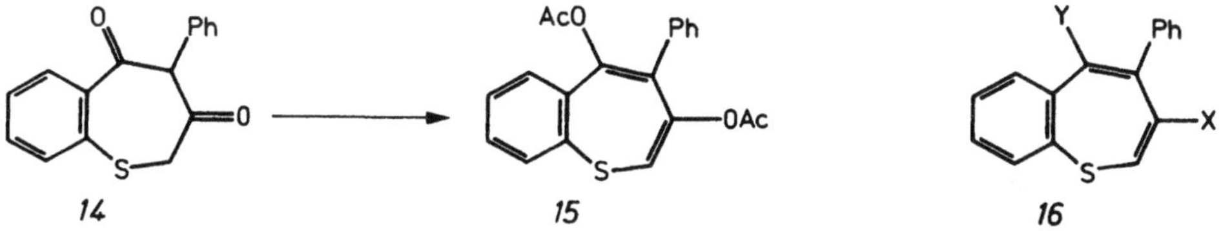

14 *15* *16*

logy various benzo[*b*]thiepin derivatives (*16*) have been synthesized so far [14]. The ease of formation of enol ethers and/or acetates at relatively low temperature is responsible for the successful synthesis of such thermally unstable thiepins.

2.3 By Ring Expansion

2.3.1 Carbonium Ion Rearrangement

This synthetic approach involves rearrangement of the incipient carbonium ion derived from the readily available six-membered ring compounds. Acid catalyzed [15, 16] and solvolysis [17] reactions of *17a* and *17b*, respectively, afforded dibenzo[*b,f*]-thiepin (*9*) which was also obtained by reaction of thioxanthylium ion (*18*) with diazomethane [18, 19].

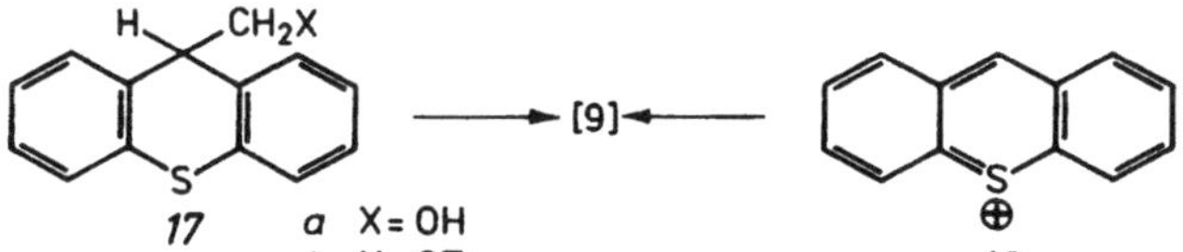

17 *a* X = OH
b X = OTs

18

2.3.2 Cleavage of Cyclopropane

The first attempted synthesis of a benzo[*b*]thiepin derivative was the solvolysis of 7,7-dichloro-3,4-benzo-2-thiabicyclo[4.1.0]heptene (*19*) [20]. Unfortunately, the reaction of *19* in hot quinoline led to 2-chloronaphthalene which suggested the reaction mechanism as shown below. In the case of the reaction of 1,1-dichloro-7b-ethoxy-

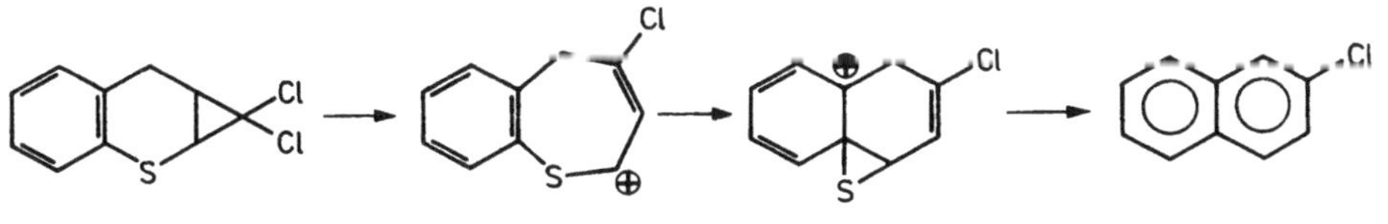

19

2-methylcyclopropa[*c*]benzo[*b*]thiapyran (*20*) with sodium methoxide in dimethyl sulfoxide, however, the evidence for the formation of the unstable benzo[*b*]thiepin (*21*) together with *22* and *23* was reported [21].

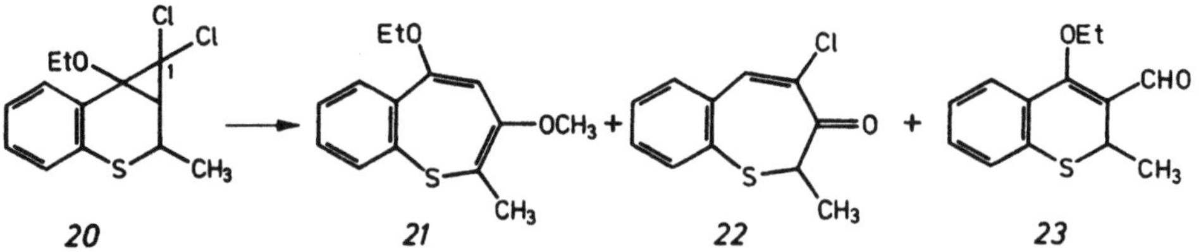

20 *21* *22* *23*

Traynelis et al. described the preparation of 4-chlorobenzo[*b*]thiepin (*24*) via the key intermediate, 7a-chlorocyclopropa[*b*]benzo[*b*]thiapyran-7-one (*25*)[6]. The ketone *25* underwent reduction with sodium borohydride to give the corresponding alcohol, which was ring opened with hydrochloric acid to yield the precursor *26* of

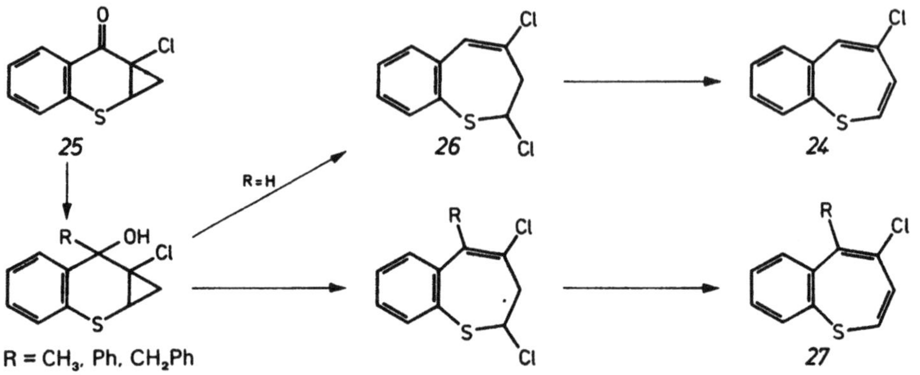

24. The same authors also observed that *25* underwent a Grignard reaction which provided a new route for synthesizing various 5-alkyl or 5-aryl derivatives of benzo[*b*]thiepin (*27*) [22].

2.3.3 Valence Isomerization of the 2-Thiabicyclo[3.2.0]heptadiene Moiety

In principle, a valence isomerization of thiabicyclo[3.2.0]heptadiene skeleton would lead to a thiepin ring system. Wynberg et al. [23] reported that the photochemical adduct (*28*) from benzo[*b*]thiophene and dimethyl acetylenedicarboxylate was not thermally stable. When heated in diglyme, it loses sulfur to give dimethyl 1,2-naphthalenedicarboxylate. This reaction presumably proceeds via ring opening of *28* to 2,3-dimethoxycarbonylbenzo[*b*]thiepin (*29*) which readily eliminates sulfur. This synthetic route was successfully applied to the reaction of electron-deficient acetylenes with enamines of 2,3-dihydrobenzo[*b*]thiophen-3-ones in which the enamine moiety constitutes part of a thiophene system. When 3-pyrrolidin-1-yl-benzo[*b*]thiophene (*30*) was allowed to react with dimethyl acetylenedicarboxylate

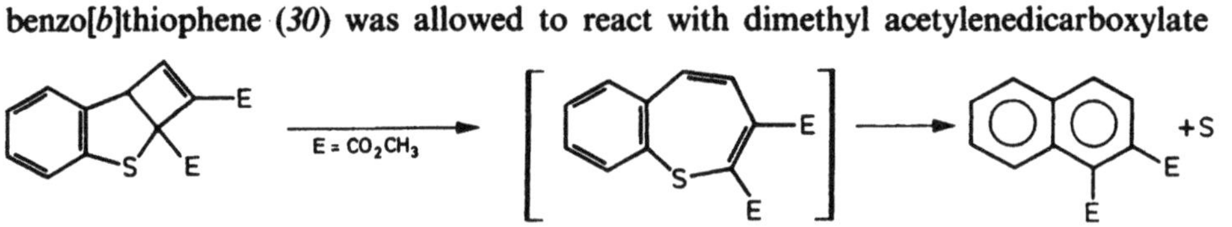

28 *29*

in ether, the benzo[*b*]thiepin (*32*) was obtained via an intermediate cycloadduct (*31*)[24,25]. The thiepin *32* was hydrolyzed under mild, acidic conditions to 3,4-

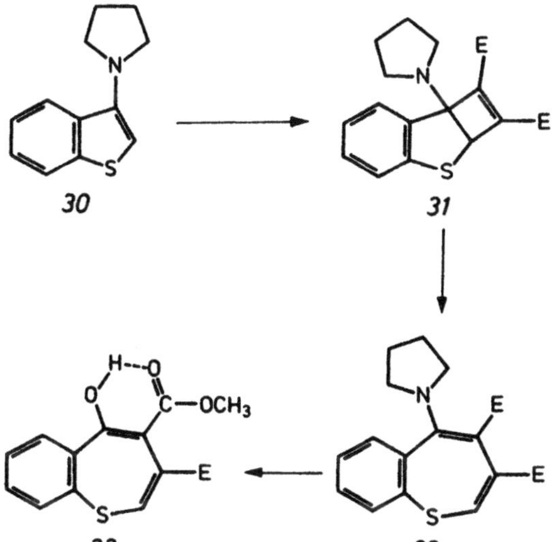

30 31 33 32

bis(methoxycarbonyl)-5-hydroxybenzo[*b*]thiepin (*33*). The existence of *33* predominantly in its enol form is attributed to the formation of a suitable hydrogen bond between the enolic hydroxy group and the methoxycarbonyl moiety. The first synthesis of a monocyclic thiepin has been achieved by the application of this synthetic method and will be discussed in Section 4.2.

2.4 By Condensation

2.4.1 Intermolecular Condensation

In the early stage of thiepin chemistry, Scott[26] reported the synthesis of benzo-[*d*]thiepin-2,4-dicarboxylic acid (*34b*) by condensation of phthalaldehyde with diethyl 3-thiapentanedioate followed by hydrolysis. Shortly after Scott's synthesis, Dimroth et al.[27] found that the ester *34a* was thermally more stable than the free acid

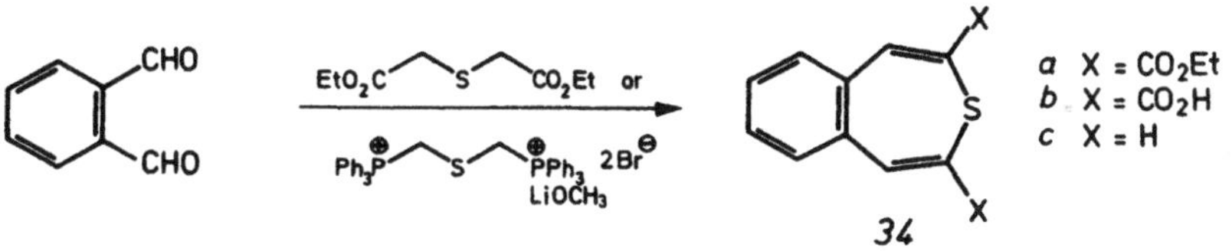

34

34b. Some related thiepins such as *35*[28], *36*[29], and *37*[30] have been synthesized utilizing the reaction of diethyl 3-thiapentanedioate with the corresponding dialde-

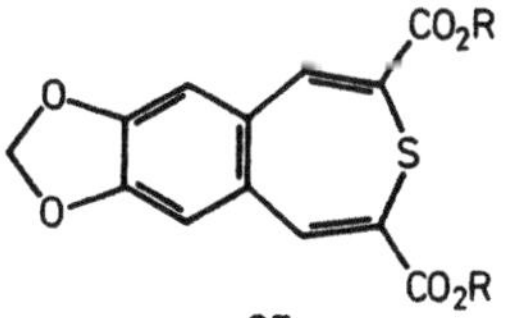

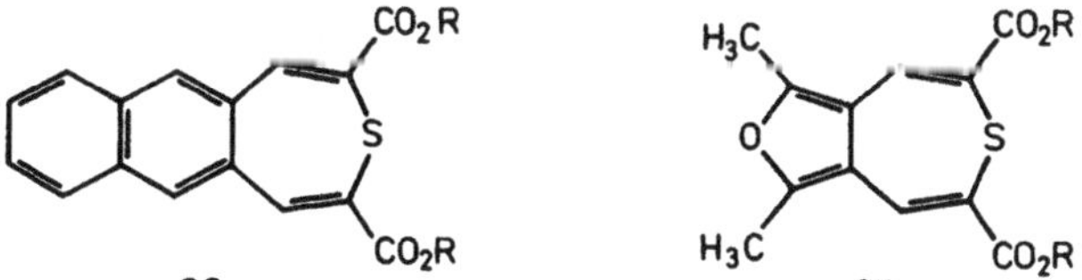

35 36 37

hydes. An elegant new preparation of the thiepin skeleton involves a double Wittig reaction. Thus, the labile parent benzo[*d*]thiepin (*34c*) has been prepared by the double Wittig reaction of phthalaldehyde with the bis-phosphonium salt prepared from α,α′-bis-bromomethyl thioether in the presence of lithium methoxide as a base at −30 °C [31]. This method has been successfully applied to the synthesis of some bridged thiaannulenes [32].

2.4.2 Intramolecular Condensation

Dibenzo[*b,f*]thiepin derivative (*39*) has been obtained by cyclization of 2-arylthio-5-nitrophenylpyruvic acid (*38*) in the presence of polyphosphoric acid [33]. The use of this Friedel-Crafts type reaction is restricted to the synthesis of the stable di- and tri-annelated thiepins such as *40* [33], *41* [7], and *42* [34].

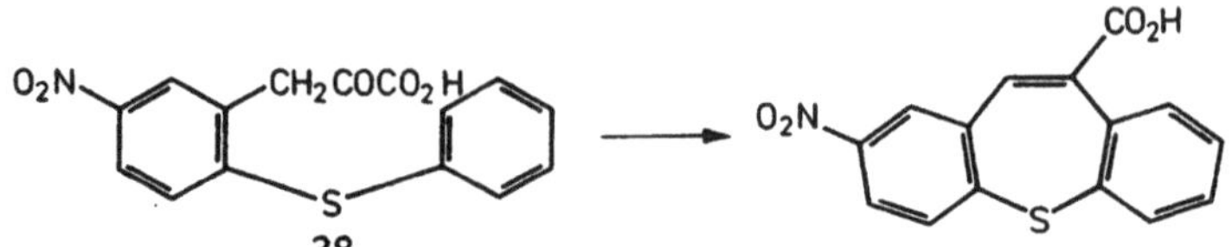

38 *39*

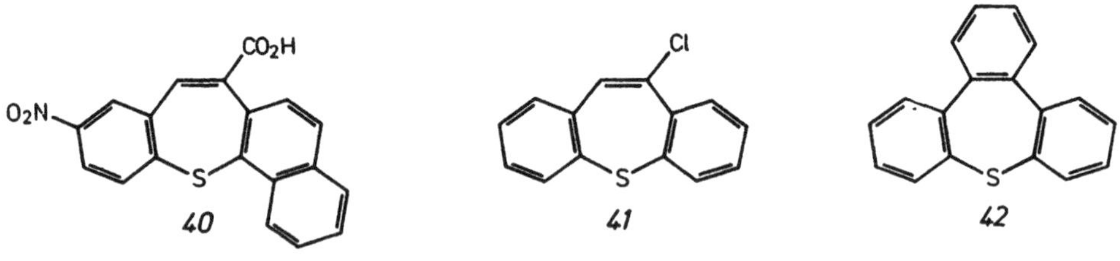

40 *41* *42*

3 New Synthetic Routes to the Thiepin System

The many efforts of synthesizing the thiepin system described in Section 2 have revealed that extremely mild conditions have to be employed for the construction of the thiepin skeleton in order to avoid thermal sulfur extrusion from the resulting thiepins. This is an especially important prerequisite for the synthetic designs aimed at obtaining simple (thermolabile) thiepin derivatives. In our own study in this field we have previously developed new versatile routes for the synthesis of the thiepin skeleton. In this section we summarize our synthetic approaches to the relatively simple thiepin derivatives.

3.1 Valence Isomerization of the 3-Thiatricyclo[4.1.0.0^{2,7}]heptene Skeleton

It is well known that bicyclo[1.1.0]butanes are easily converted into the corresponding 1,3-butadienes under extremely mild conditions by treatment with various transition metal catalysts [35]. In fact, treatment of naphtho[1,8]tricyclo[4.1.0.0^{2,7}]heptene (*43*) with a catalytic amount of Ag(I) at 0 °C afforded pleiadiene (*44*) in good yield [36]. With these precedents in mind, bicyclobutanes incorporated in properly designed cyclic systems seem to be good candidates for the synthesis of the elusive cyclic conjugated systems.

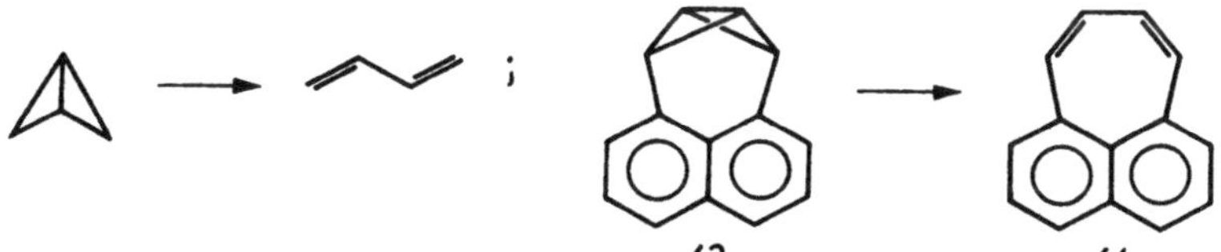

43 *44*

Thus, 4,5-benzo-3-thiatricyclo[4.1.0.0^{2,7}]heptene (*46*), which is regarded as a valence isomer of benzo[*b*]thiepin (*4*), appears to be an attractive precursor to *4*. The most obvious method for the synthesis of *46* would be the application of the elegant Katz method [37] to the benzvalene synthesis. The well established success of the preparation of *43* [36] by the Katz method makes this a likely route to *46*.

The reaction of 2*H*-benzo[*b*]thiapyran (*45*) with *n*-butyllithium in hexane at —20 °C followed by treatment with dichloromethane at —110 °C provided a 25% yield of the desired *46* as stable crystals [38]. The structure of *46* was confirmed by NMR spectral data. The ^{1}H-NMR spectrum of *46* exhibits four well separated signals at δ 1.90 (t, 2 H, J = 2.8 Hz, H-1,7), 3.04 (dt, 1 H, J = 4.0, 2.8 Hz, H-6),

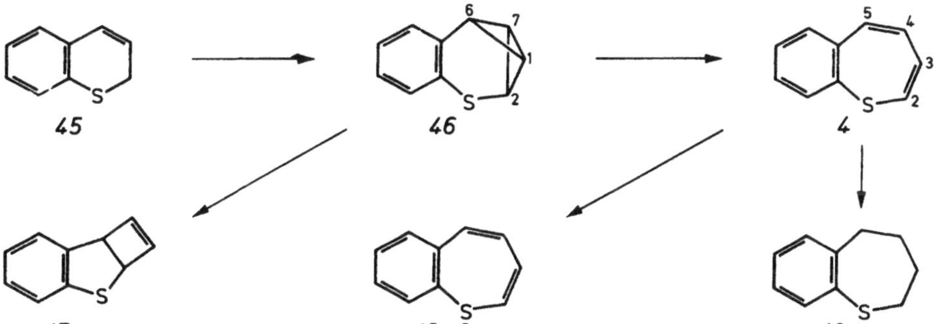

45 *46* *4* *47* *49* O$_2$ *48*

3.58 (dt, 1 H, J = 4.0, 2.8 Hz, H-2), and 6.80 (m, 4 H, aromatic). The fairly large, long-range interaction (J = 4.0 Hz) between H-2 and H-6 is characteristic of the wing protons in a bicyclobutane skeleton [39]. In the ^{13}C-NMR spectrum of *46*, the bicyclobutane carbons give peaks at δ —3.6 (C-1, 7, J_{CH} = 210 Hz), 42.3 (C-6, J_{CH} = 158 Hz), and 44.4 (C-2, J_{CH} = 172 Hz). One of the most conspicuous features is the large ^{13}C-H coupling constant observed for C-1 and C-7. This reflects the 42% s-character of the bonding carbon orbitals of C1-H and C7-H bonds. Final structural proof was obtained by X-ray analysis [40].

Although the Ag(I)-catalyzed reaction of *46* leads exclusively to the cyclobutene isomer (*47*), the conversion of *46* into benzo[*b*]thiepin has been realized by a rhodium catalyst [41]. Treatment of *46* with a catalytic amount of dicarbonyl-2,4-pentanedionatorhodium(I) in chloroform at 0 °C readily gives 57% of *4* as pale yellow needles of mp 23.5–24.5 °C, which was purified by low-temperature column chromatography on alumina at —20 °C. All manipulations during work-up have to be carried out below —10 °C in order to avoid thermal sulfur extrusion of *4*. The structure of *4* was confirmed spectroscopically and chemically. The ^{1}H-NMR spectrum of *4* measured at —20 °C (δ 5.90 (H-2), 6.34–6.56 (H-3, 4). 7.09 (H-5), and 7.15–7.50 (aromatic)) was assigned on the basis of relative chemical shifts, the shielding of H-2 being larger than that of H-3, and through comparison with the spectrum of 3,4-d_2-benzo[*b*]-thiepin [41]. Catalytic reduction of *4* yielded the known tetrahydrobenzo[*b*]thiepin (*48*). Heating of *4* in carbon tetrachloride led to the formation of naphthalene and sulfur. Treatment of *4* with *m*-chloroperbenzoic acid in dichloromethane at —20 °C afforded benzo[*b*]thiepin 1,1-dioxide (*49*). Irradiation of *4* in tetrahydrofuran with a 450 w high-pressure mercury lamp through a pyrex filter at 0 °C yielded the valence isomer (*47*) along with a small amount of naphthalene.

This unique method could be applied to the synthesis of some methyl, formyl, and methoxycarbonyl substituted benzo[*b*]thiepins [42,43,44]. Since the bridgehead proton abstraction from a bicyclobutane by *n*-butyllithium is a favored process, methyl groups can be introduced into the 1- and 7-positions of *46* prior to its transformation into the thiepin. Treatment of *46* with an equivalent of *n*-butyllithium in tetrahydrofuran at —60 °C followed by quenching of the resulting anion with methyl iodide gave the 7-methyl derivative (*50*) which was quantitatively isomerized to a 1:1.3 mixture of 3-methyl- (*51*) and 4-methylbenzo[*b*]thiepin (*52*) by treatment with the rhodium catalyst [44]. The separation of *51* and *52* was achieved by column chromatography on alumina at —20 °C. For the synthesis

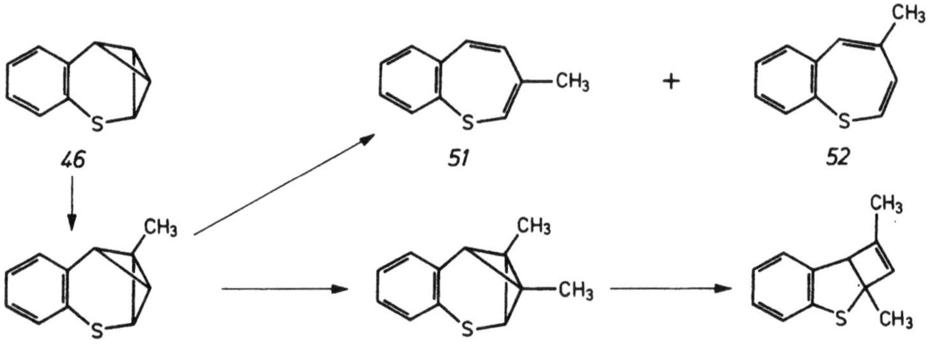

of 3,4-dimethylbenzo[*b*]thiepin, compound *53* was prepared by successive methylation of *50*. When treated with the rhodium catalyst, however, the dimethylbicyclobutane derivative *53* exhibited an exclusive rearrangement to the cyclobutene isomer (*54*).

The synthesis of the "peri" methyl substituted benzo[*b*]thiepins, *58* and *59*, was conveniently accomplished via the corresponding bicyclobutane derivatives *56* and *57* starting from 4-methyl-2 H-thiachromene (*55*).

The syntheses of 4-methoxycarbonyl- (*60*) and 4-formylbenzo[*b*]thiepin (*61*) were also achieved by a procedure involving conversion of the 1-lithio derivative of *46* to the corresponding bicyclobutanes by treatment with methyl chloroformate and ethyl

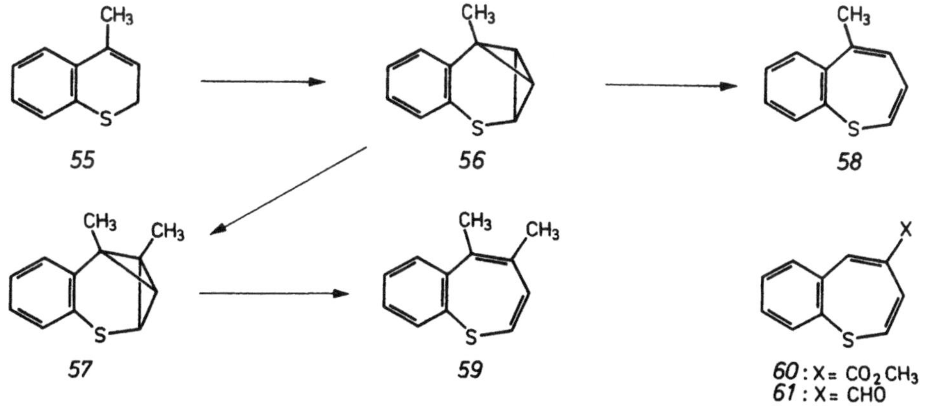

formate, respectively, and isomerization of the resulting compounds with the rhodium catalyst. Unfortunately, owing to the high rate of sulfur extrusion, *60* and *61* were contaminated with the respective naphthalene derivative. However, pure 4-ethoxycarbonylbenzo[*b*]thiepin can be obtained by the alternate synthetic route described in Section 3.2.

Obviously, the advantages of this new synthetic method are i) valence isomerization can be performed under very mild conditions and ii) due to the fairly strong acidity of the bridgehead hydrogens of a bicyclobutane moiety, some substituents can be introduced regioselectively. Nevertheless, one serious disadvantage of this synthesis is the limitation of its applicability. Thus, the formation of the precursor bicyclobutane derivatives is limited only to *46* and *56*, and many attempts to prepare bicyclobutane precursors from other thiapyran derivatives have been unsuccessful so far.

3.2 Rearrangement via Carbene Intermediates

Although the merits of the synthetic method described in the preceding section are apparent, the limited applicability of this method prompted the search for alternative synthetic routes. During our work in this field we have found that a well known ring expansion via a carbene intermediate [45] does indeed proceed remarkably well with a variety of precursors.

The readily accessible dibenzothiapyrylium salt (*62*) [46] reacts with ethyl lithio diazoacetate [47] in a 1 : 1 mixture of ether and tetrahydrofuran at −120 °C to form the diazo compound (*63*). Treatment of *63* with 5 mol-% of π-allylpalladium chloride dimer in a 1 : 2 mixture of chloroform and carbon tetrachloride at 0 °C and

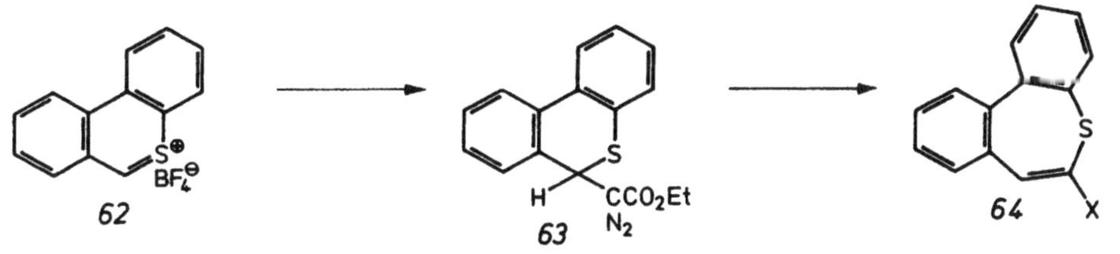

allowing the mixture to warm to room temperature for 2 h affords quantitatively the desired 6-ethoxycarbonyldibenzo[*bd*]thiepin (*64*: X = COOEt)[48]. Alkaline hydrolysis of *64* (X = COOEt) gives the free acid which, on heating with copper in quinoline at 150 °C for 30 min, gives the parent dibenzo[*bd*]thiepin (*64*: X = H). Despite the presence of the thiepin skeleton, *64* (X = H) is a stable compound. However, on thermolysis in refluxing xylene in the presence of triphenylphosphine for 1 day, the thiepin is converted into phenanthrene.

The synthetic utility of the ring expansion reaction was demonstrated by its application to the synthesis of thermolabile thiepins. When the diazo compound (*66*) obtained from benzo[*c*]thiopyrylium salt *65* was treated with palladium catalyst under the same conditions as in the case of *63*, the product isolated was ethyl 2-naphthoate (*68*)[48]. The plausible reaction pathway is one comprising i) decomposition of *66* to the corresponding carbene intermediate, ii) ring expansion to the

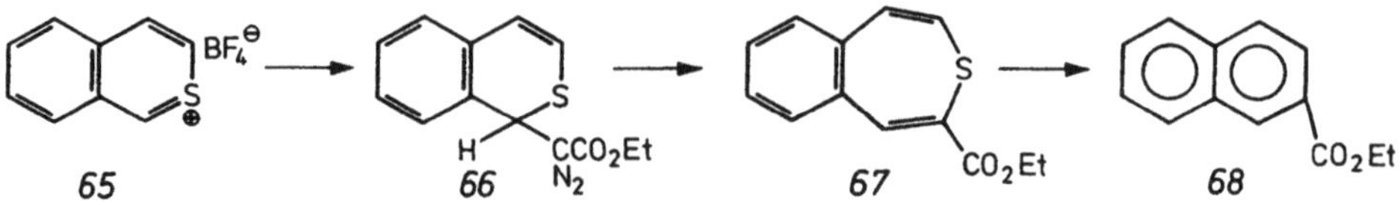

65 *66* *67* *68*

thiepin *67*, iii) desulfurization of *67* at room temperature. In order to gain information about the pertinent reaction conditions leading to the thiepin *67*, we subsequently lowered the reaction temperature to —60 °C and monitored the course of the reaction by ^{1}H-NMR spectroscopy[49] as the solution was gradually heated. The ^{1}H-NMR spectrum of *66*, immediately after mixing the sample with the palladium catalyst at —60 °C exhibited only signals of *66* contaminated with the small signals attributed to *67* at δ 7.90 (s) and 6.01 and 6.84 (AB-q). The spectrum of this solution allowed to heat to —50 °C for 90 min clearly indicated a half conversion of *66* into *67*. The almost complete disappearance of *66* accompanied by the exclusive formation of *67* was reached at —25 °C corresponding to a period of 150 min after the start of the reaction. Although a slight amount of *68* was formed at 0 °C (after 180 min from the start), most of the signals of the thiepin *67* still remained unchanged. Finally, considerably amounts of ethyl 2-naphthoate (*68*) were produces at 30 °C. The isolation of *67* could also be achieved. Treatment of *66* with the palladium catalyst at —10 °C for 2 h followed by low-temperature column chromatography on silica gel at —40 °C provided pure *67* as a yellow liquid. The exomethylene compound (*69*) was also formed as a by-product in 7% yield.

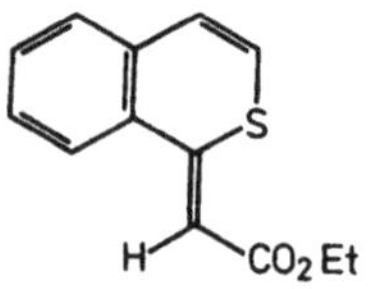

69

A series of ethoxycarbonylbenzo[*b*]thiepins (*73*–*76*) has also been synthesized by this method[50]. When benzothiopyrylium salt (*70*) was treated with ethyl lithiodiazoacetate in a mixture of ether and tetrahydrofuran at —120 to —110 °C, the expected diazo compounds (*71*) and (*72*) were obtained. Treatment of *71* with palladium catalyst (2 mol-%) in chloroform at —60 °C afforded 2- (*73*) and 3-ethoxycarbonylbenzo[*b*]thiepin (*74*) in yields of 27% and 19%, respectively. In exactly the same procedure, the alternative diazo compound (*72*) also underwent rearrangement to a mixture which was separated into 4- (*75*) and 5-ethoxycarbonylbenzo[*b*]-thiepin (*76*) in 31 and 15% yield, respectively. Due to the competitive hydrogen

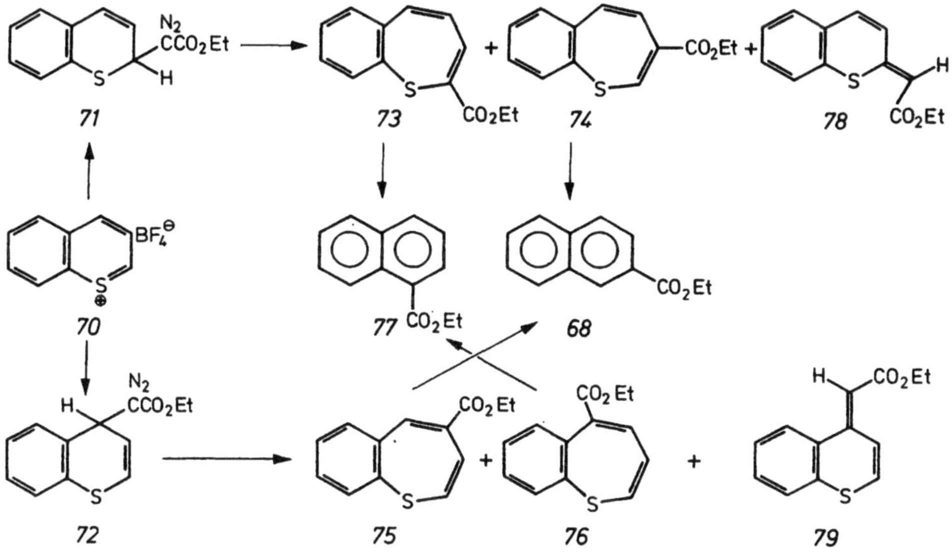

shift to the intermediate carbenes, separation of the mixture of the above reaction products from *71* and *72* led to isolation of small amounts of the exo-methylene compounds *78* and *79*, respectively. The structures of all four thiepins are unambiguously established by consistent ^{1}H-NMR spectra obtained at —30 °C in

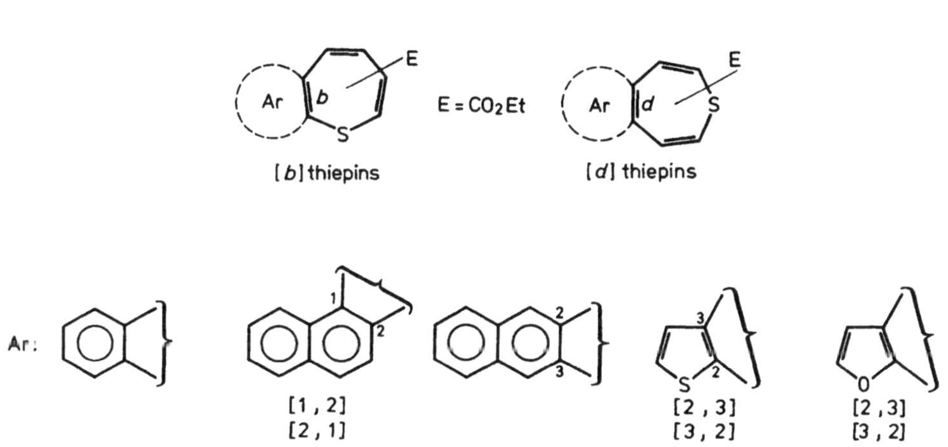

Scheme 1. Annelated [*b*]- and [*d*]thiepins synthesized by a carbene route

addition to the ready quantitative conversion of these thiepins to either ethyl 1-naphthoate (*77*) or ethyl 2-naphthoate (*68*) at room temperature.

The synthetic method described in this section can be successfully applied not only to the benzo-annelated thiepins but also to the various kinds of annelated and non-annelated thiepin derivatives. The annelated thiepins thus obtained, though some of them were not isolated, are shown in Scheme 1 [49]. The thermal stability of these thiepins can be evaluated by their half-lives at an appropriate temperature measured by ^{1}H-NMR spectroscopy and will be discussed in Section 5.1.

The most successful application of this synthetic route is the first synthesis and isolation of a simple and stable monocyclic thiepin which will also be described in Section 4.4.

4 Synthesis of Monocyclic Thiepins

Thermally stable thiepins have been synthesized in the form of annelation with aromatic ring(s), but the pronounced annelated nature of these thiepins vitiates their usefulness for detailed studies as 8π-electron systems. In the study of thiepins the synthesis of non-annelated thiepins is therefore very important. At the present time of writing this article some monocyclic thiepins have been synthesized and characterized.

4.1 Stabilizing Device for the Thiepin Skeleton

Most of the many attempts to prepare monocyclic thiepins have been unsuccessful. This is due to the fact that thiepins readily extrude sulfur and are transformed into the corresponding benzene derivatives.

In general, two factors relating to the stabilization of the thiepin skeleton might be considered.

Firstly, the electronic effect of the substituents would be a dominant factor in the determination of the thermal stability of the thiepins. Thermal stability was found to decrease sharply with decreasing substitution. For example, the tri-substituted benzo[*b*]thiepin (*16d*) extrudes sulfur only when heated while the non-substituted benzo[*b*]-thiepin (*4*) is readily converted into naphthalene and sulfur at 40 °C with a half-life of 80 min [41].

Secondly, in view of the generally accepted mechanism for the sulfur extrusion reaction, which involves valence isomerization of the thiepin ring to its corresponding thianorcaradiene (benzene episulfide) isomer followed by irreversible loss of sulfur,

Scheme 2. Steric effect of two bulky groups on the thiepin-thianorcaradiene equilibrium

it might be feasible that a simple thiepin can be prepared if the energy difference between the ground state of the thiepin form and the thianorcaradiene form were reasonably large. Molecular models of a thiepin possessing two bulky groups at the 2- and 7-positions indicate that these groups force the non-bonding interactions in the corresponding thianorcaradiene structure to be large, and hence the thiepin form will be favored as illustrated in Scheme 2. This concept has been supported by the successful isolation of some stable monocyclic thiepins which will be discussed in Sections 4.3 and 4.4.

4.2 Reinhoudt Synthesis

In 1972 Wynberg and Helder [51] reported a facile reaction between tetramethylthiophene (*80*) and dicyanoacetylene at room temperature in the presence of aluminum chloride. On the basis of spectroscopic data of the reaction product as well as its sulfone, 3,4-dicyano-2,5,6,7-tetramethylthiepin structure (*81*) was assigned. The product was pyrolyzed at 140 and 300 °C to *83* and *84*, respectively. Un-

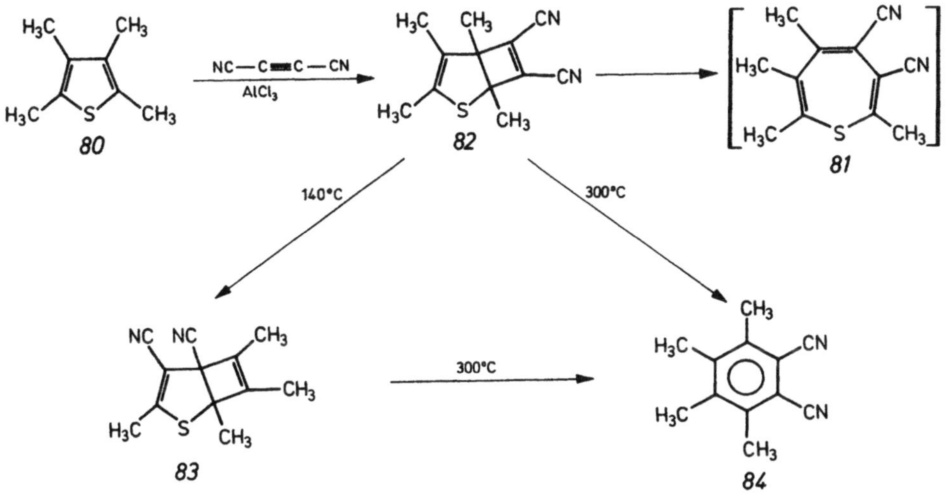

fortunately, however, the assignment of a thiepin structure (*81*) to the product of the above reaction has been shown to be incorrect [52]. ^{13}C-NMR spectral data show it to be a 2-thiabicyclo[3.2.0]heptadiene (*82*) formed by [2 + 2]cycloaddition of *80* to dicyanoacetylene. Thermal isomerization of *82* to *83* can be interpreted as Cope rearrangement which is the first example of a symmetry-allowed thermal [3.3]antara, antara sigmatropic shift of a 2-heterobicyclo[3.2.0]heptadiene system [52].

Reinhoudt et al. [53] have reported the first synthesis of a monocyclic thiepin stabilized by electronic effects of the substituents. This synthesis utilizes the idea described in Section 2.3.3. 3-Methyl-4-pyrrolidinothiophene (*85a*) was treated in deuteriochloroform at −30 °C with dimethyl acetylenedicarboxylate. ^{1}H-NMR monitoring of the reaction indicated that a [2 + 2]cycloaddition proceeded slowly at this temperature giving the 2-thiabicyclo[3.2.0]heptadiene (*86a*) which rearranged via ring opening of the cyclobutene moiety to the 4-pyrrolydinylthiepin (*87a*). At the reaction temperature, however, sulfur is slowly extruded from *87a* to give a benzene derivative (*88a*). At —30 °C, the rates of cycloaddition, ring expansion and sulfur

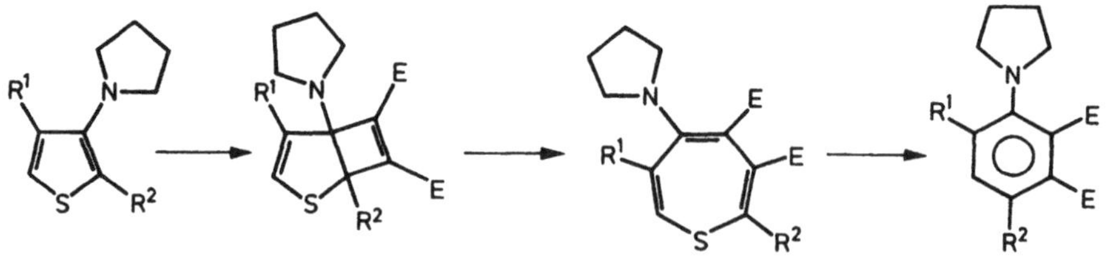

85 86 87 88

a: $R^1 = CH_3$, $R^2 = H$ *b*: $R^1 = R^2 = H$ *c*: $R^1 = H$, $R^2 = CH_3$

extrusion are of the same order of magnitude, and the optimum yield of *87a* was reached after 80 h. Similarly, 3-pyrrolidinothiophene (*85b*) reacted with dimethyl acetylenedicarbocylate to yield the corresponding thiepin (*87b*). On the other hand, *85c* reacted with dimethyl acetylene dicarboxylate only at room temperature, and it was impossible to observe the thiepin (*87c*) under these conditions.

4.3 Schlessinger Synthesis

The stabilizing effect of sterically bulky groups at the 2- and 7-positions of the thiepin ring has been demonstrated experimentally by Schlessinger et al. [30]. Treatment of furo[3,4-*d*]thiepin (*89*) or its S-oxide *90* with N-phenylmaleimide (NPM) at 25 °C yields a 3:2 mixture of the exo- and endo-adducts *91* with the elimination of sulfur [54]. The 2,4-bis-methoxycarbonyl substituted furothiepin *92a* reacts similarly. Thus, treatment of *92a* with N-phenylmaleimide at 120 °C gave a 7:3 mixture of the

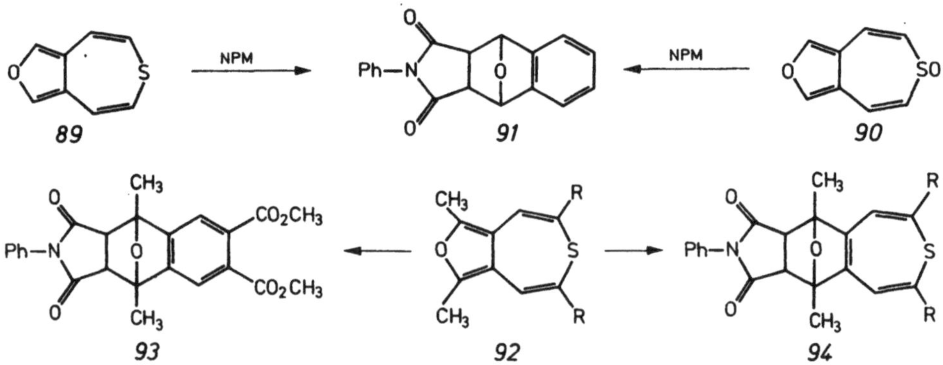

a: $R = CO_2CH_3$ *b*: $R = C(CH_3)_2OH$

endo- and exo-adducts *93* [54]. The great reactivity of these furanothiepins suggested that by introduction of bulky groups *R* into *92* it might be possible to prepare a stable non-annelated thiepin. To this end, the diester *92a* was converted into the diol (*92b*) which can be regarded as a suitable precursor for the synthesis of *94*. As expected, when *92b* was treated with N-phenylmaleimide in dichloromethane, the desired thiepin *94* was obtained as a highly crystalline compound [30]. The thiepin *94* as bright yellow crystals melted at 180 °C with decomposition. Although *94* is a stable compound which cannot be transformed into *93* even upon prolonged heating in dimethylformamide or chloroform, thermolysis in the presence of triphenylphosphine rapidly furnishes *93*.

It should be noted that although the thiepin *94* exhibits a tetracyclic structure, it is the first example of an isolable 8π electron thiepin.

4.4 Murata Synthesis

Our study aimed at monocyclic thiepin synthesis is based on the finding that the unusual thermal stability of Schlessinger's thiepin *94* may originate from the presence of two bulky groups at the 2- and 7-position of the thiepin ring. Therefore, our hope was to synthesize a simple and stable monocyclic thiepin via a carbene intermediate. At first, 2,7-diisopropyl-4-ethoxycarbonylthiepin (*96*) was chosen as our target. 2,6-Diisopropylthiopyrylium salt *95*, a key intermediate in the synthesis of *96*, was obtained by the stepwise synthesis starting from 3-bromo-4-methylpentan-2-one and methyl-2-mercapto-3-methylbutanoate [55]. Reaction of *95* with ethyl lithiodiazoacetate in a 3:1 mixture of tetrahydrofuran and ether at −110 °C furnishes a single diazo compound (*97*) in 94% yield. Although *97* seems to be a precursor for the synthesis of *96*, treatment of *97* with palladium catalyst at room temperature affords 2,5-diisopropyl-4-ethoxycarbonylmethylene-4 *H*-thiapyran (*99*) which is formed from the carbene intermediate *98* via hydrogen shift [55]. In order to prevent hydrogen shift, a methyl group was introduced into the 4-position of *95*. Treatment of *95* with methyllithium yielded 2,6-diisopropyl-4-methyl-4 *H*-

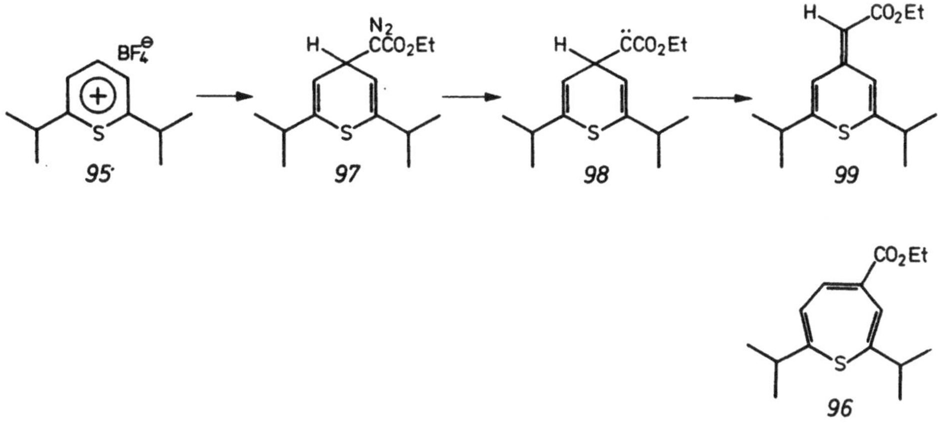

thiapyran which was converted into the corresponding diazo compound *100* in the usual way. The reaction of *100* with palladium catalyst was monitored by ^{1}H-NMR spectroscopy. At −110 °C only signals attributable to *100* were observed. Warming

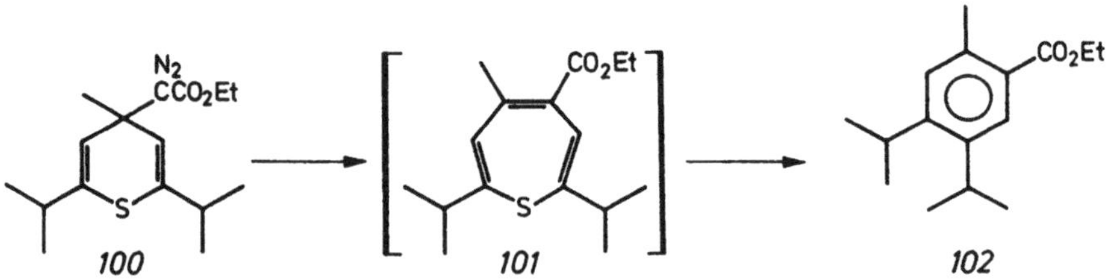

100 101 102

the reaction mixture to —70 °C for 30 min led to a gradual decrease of these signals due to the presence of *100* and the appearance of new signals which remained almost unchanged for at least 1.5 h at this temperature, but gradually increased at —50 °C during 0.5 h. All the signals caused by *100* finally replaced by a new set of signals which were consistent with the structure of ethyl 4,5-diisopropyl-2-methylbenzoate (*102*) at room temperature. Work-up of the reaction mixture led to isolation of pure *102* in 70% yield. While the formation of *102* is strong evidence of the intermediacy of the thiepin *101*, no NMR spectroscopic information on the formation of *101* was obtained. These results suggested that the thiepin *101* is unstable even at —70 °C. Therefore, it is concluded that the stabilizing effect of the bulky isopropyl groups at C-2 and C-7 of a thiepin ring is not sufficient to permit isolation of *101*.

Since the steric requirement of the *tert*-butyl group was known to be much larger than that of the isopropyl group, as exemplified by the free energy difference involved in the axial-equatorial equilibrium of substituted cyclohexanes [56], 2,7-di-*tert*-butylthiepin derivatives seemed to be more promising candidates for the synthesis.

The precursor of our thiepin synthesis, 2,6-di-*tert*-butyl-4-methylthiopyrylium salt (*103*), can be prepared either from 2-methylthiopyrylium salt via a sequence of reactions [i, *t*-BuMgCl; ii, $Ph_3C^+BF_4^-$; iii, *t*-BuLi; iv, $Ph_3C^+BF_4^-$] or from 2-*tert*-butylthiophene via a sequence of reactions [i, *t*-BuCOCl/$SnCl_4$; ii, Li/NH_3/*t*-BuOH; iii, Zn/$(CH_3)_3SiCl$; iv, $Ph_3C^+BF_4^-$; v, CH_3Li; vi, $Ph_3C^+BF_4^-$]. Treatment of the diazo compound *104* obtained from *103* with palladium catalyst at 0 °C gave the desired thiepin *105* as yellow prisms of mp 23.5–24.5 °C in almost quantitative yield [57]. The structure of *105* was determined by elemental analysis and spectral

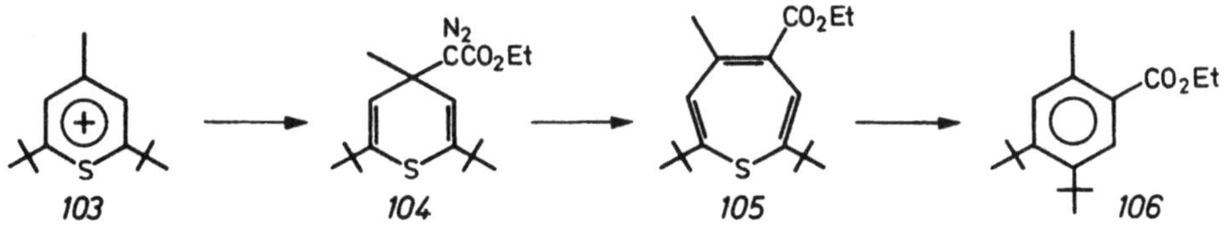

103 104 105 106

data and conversion of *105* into the corresponding benzene derivative *106*. In spite of its monocyclic thiepin structure, *105* shows remarkable thermal stability and can be handled under usual conditions with no detectable decomposition. However, on prolonged heating in toluene at 140 °C in a sealed tube *105* was converted in nearly quantitative yield into *106* and sulfur. A comparison of *105* and *101* shows that the substitution of the *tert*-butyl groups for isopropyl groups at C-2 and C-7 of the thiepin ring produces high thermal stability. Presumably, the thianorcaradiene intermediate is not formed owing to increased steric hindrance between two *tert*-butyl groups.

The study on 2,7-di-*tert*-butylthiepin has recently been extended to explore more simply substituted thiepins [58]. The palladium-catalyzed reaction of the diazo compound *107* lacking a 4-methyl substituent gives exclusively the exo-methylene compound *108* whereas the acid-catalyzed reaction of the same precursor *107* resulted in the formation of 2,7-di-*tert*-butyl-4-ethoxycarbonylthiepin (*109*) [58]. Due to the substantial thermal stability of *109* it is possible to transform the ethoxycarbonyl group into the hydroxymethyl (*110*), trimethylsilyloxymethyl (*111*) and formyl group (*112*) [58].

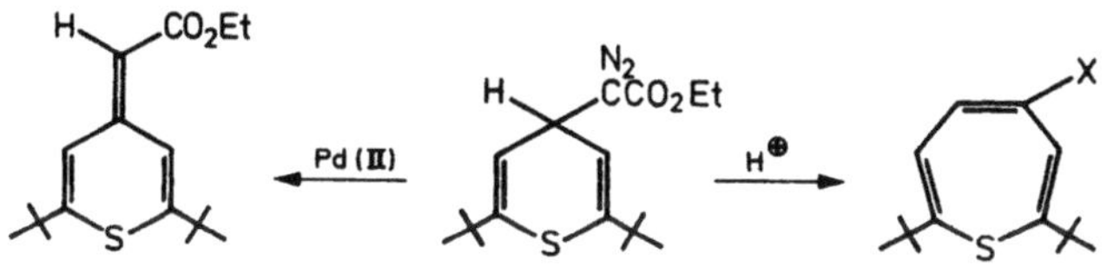

108 *107*

109 : X = CO_2Et
110 : X = CH_2OH
111 : X = CH_2OTMS
112 : X = CHO

5 Thermal Stability

Through the recent developements in thiepin chemistry, many thiepin derivatives are now available for a detailed study of their properties. One of the most intriguing problems awaiting solution is the structural effects on the thermal stability of the thiepin system.

5.1 Electronic Effects of Substituents

The synthetic procedures outlined in Section 3 provide ready access to a wide variety of benzo[*b*]thiepins in which the substituents are methyl, formyl and ethoxycarbonyl. The thermal stability of the thiepins was conveniently evaluated from the half-lives of their sulfur extrusion reactions by means of ^{1}H-NMR spectroscopy. The half-lives of a series of substituted benzo[*b*]thiepins and naphtho[2,3-*b*]thiepins thus obtained are summarized in Table 1.

Recently, Reinhoudt and Kouwenhoven [53] have reported, in connection with their successful synthesis of monocyclic thiepin *87*, that the relatively high stability of *87* is attributed to the presence of the two methoxycarbonyl groups which cause a decrease in the electron density of the 8π-electron system. As a result, the formal anti-aromatic character is reduced. Traynelis and his coworkers [6] have also reported a slight increase in the thermal stability of benzo[*b*]thiepin when electron-withdrawing groups are present. These conclusions have subsequently been supported by resonance energy calculations on various thiepin derivatives [59]. Thus, the re-

Table 1. Half-lives of some benzo[*b*]- and naphtho[2,3-*b*]thiepins

Compound	X	Temp.	Half-lives [min]				
		[°C]	3-X	4-X	5-X	parent	2-X
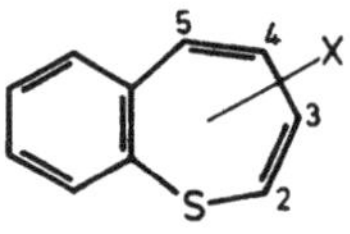	COOEt	27	16				
	COOEt	37		58	69	182	
	COOEt	47				58	352
	CHO	47		49		58	
	CH_3	47	100	62	67	58	
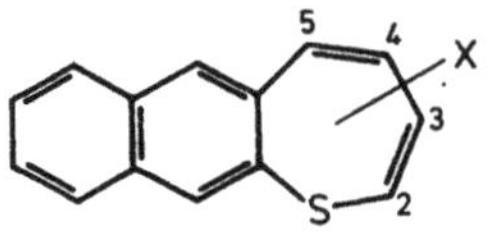	COOEt		159 min at 74 °C	60 min at 133 °C			41 min at 146 °C

sonance energies of 430 substituted thiepins have been calculated using a simple Hückel method, and the results predict that thiepins substituted by electron-withdrawing groups such as methoxycarbonyl and fluorine will be less antiaromatic than the parent thiepin [59].

However, inspection of Table 1 shows that, in sharp contrast with the above arguments, benzo[*b*]thiepin is stabilized by the electron-donating methyl group whereas the reverse is true for π-accepting formyl and/or ethoxycarbonyl groups, except 2-substituted thiepins. It should be emphasized that the stabilizing and destabilizing effects caused by methyl and ethoxycarbonyl groups, respectively, are most prominent when these groups are introduced into C-3. Furthermore, in both series of benzo[*b*]thiepins, a comparable stability is found between 4- and 5-substituted derivatives.

Recently, Traynelis and his coworkers [22] have reported the thermal decomposition of some substituted benzo[*b*]thiepins (Table 2). A slight increase in the stability of the thiepin ring by phenyl, benzyl, methyl, chloro, and bromo substituent was

Table 2. Approximate half-lives of some benzo[*b*]thiepins

Substituent	Half-life [h]	
	at room temp	at 30 °C
none	17	
4-Cl	22	12
2-Cl	42	
4-Cl, 5-Ph	27	17
4-Cl, 5-CH_2Ph	46	22
4-Br, 5-Ph	47	
4-Cl, 5-CH_3		25
4-Br, 5-CH_3		29

suggested. In addition, larger stabilization is achieved by groups with greater electron-donating ability. This trend supports the recently reported observations [50]. Therefore, the relatively high stability of the thiepin *87* must be attributed to factors other than the presence of electron-withdrawing methoxycarbonyl groups.

Although an explanation for the substituent effects on the thermal stability of the thiepin ring awaits further systematic studies, an attractive way of explaining these effects would be a concept of "effective descontinuity of cyclic conjugation" proposed recently by Inagaki [60].

5.2 Annelation Effects of Aromatic Rings

It is well known that unstable molecules can be stabilized by varying the aromatic ring annelation. This is also the case for thiepins. The rise in thermal stability of the thiepin ring with increasing number of annelated benzene rings is seen from the following comparison. While the unsubstituted parent thiepin *1* is considered to be an extremely reactive species and could not be detected so far, benzo[*b*]thiepin (*4*) has been

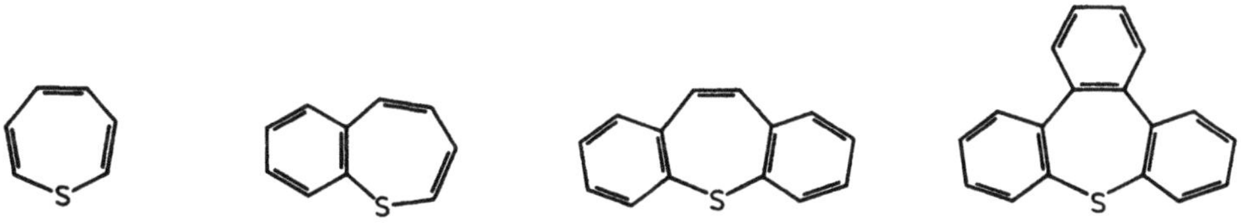

1 unknown; *4* $t_{1/2}$ = 58min/47°C; *9* stable; *42* stable

isolated, though very thermolabile, in crystalline form. The dibenzothiepins *9* and *64* are found to be stable compounds and the latter, on thermolysis in refluxing xylene in the presence of triphenylphosphine for 1 day, furnishes phenanthrene [17]. The thermal stability of the tribenzothiepin *42* is apparent from its synthesis which involves cuprous chloride-catalyzed thermal cyclization of 2-phenyl-2′-sulfonylchlorodiphenyl sulfide at 250–260 °C [34]. Furthermore, the thiepin *42* was readily converted into triphenylene after 5 h at 380 °C in a mixture of bisphenoxyterphenyls in the presence of copper bronze. The characteristic feature of the thiepin skeleton, however, decreases sharply with increasing number of annelated benzene rings. For example, the ultraviolet spectrum of *42* resembles that of *o*-terphenyl indicating the absence of the major influence of the 8π-electron thiepin ring [34].

Table 3. Effect of annelation site of benzene and naphthalene on the thermal stability of thiepins

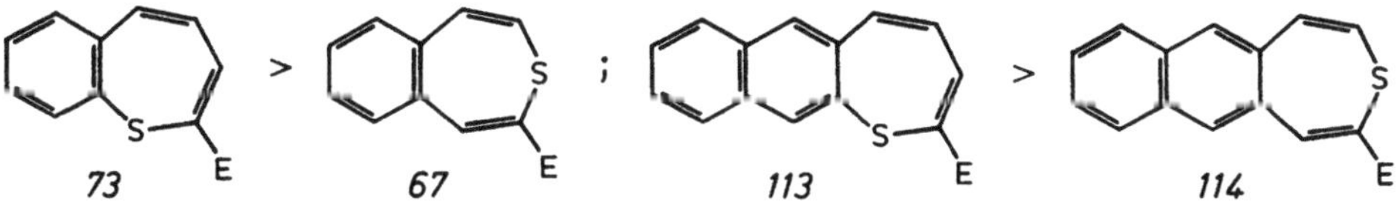

73 > *67* ; *113* > *114*

E = CO_2Et

73: $t_{1/2}$ = 352 min/47°C; *67*: 57 min/25°C; *113*: 49 min/146°C; *114*: 72 min/94°C

Now, we would like to focus our attention on the site of the thiepin ring annelated with aromatic rings. The two series of thiepins shown in Table 3 illustrate this effect. It is apparent from the comparison that the benzo[*b*]thiepin *73* and the naphtho[2,3-*b*]thiepin *113* are more stable than the corresponding [*d*]fused thiepins, *67* and *114*, respectively.

Our recent synthesis of various types of annelated thiepins described in Section 3.2 makes it possible to compare the effects of aromatic and heteroaromatic rings on the stability of the thiepin skeleton. Table 4 indicates the half-lives of a series of thiepins [*b*]fused with an aromatic ring. Clearly, the thermal stability of these thiepins increases with increasing aromatic character of the annelated counterparts [61]. Strictly speaking, the stability of thiepins is largely dependent on the Hückel bond orders of the aromatic bonds being annelated with the thiepin moiety. Thus, the increase in the thermal stability is in good agreement with the decrease in the bond order. Extreme examples of this kind of effect can be seen in the stable furo[3,4-*d*]-thiepin (*13*, X = O) and thieno[3,4-*d*]thiepin (*13*, X = S) in which the bond orders considered are 0.61 and 0.63, respectively. In this connection, it should be noted that the relatively high stability of the monocyclic thiepin *87* might in part be ascribed to the presence of a resonance contributor such as *87a* arising from the "push-pull" effect of the adjacent pyrrolidinyl and methoxycarbonyl groups as a result of which

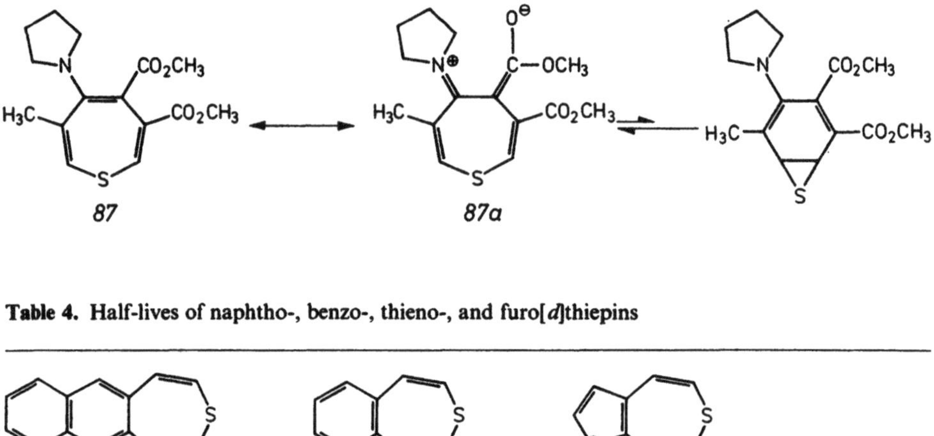

87 *87a*

Table 4. Half-lives of naphtho-, benzo-, thieno-, and furo[*d*]thiepins

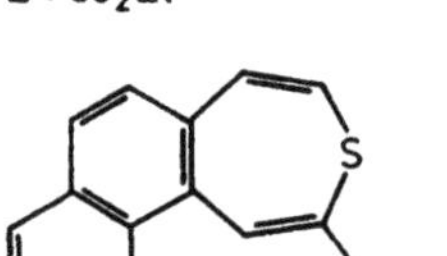

114
$t_{1/2}$ = 72 min / 94°C
P_{ij} = 0.60

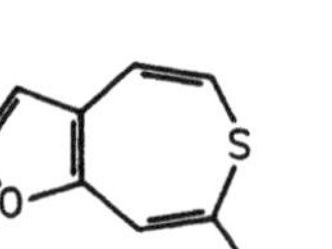

69
57 min / 25°C
0.67

115
52 min / −27°C
0.71

E = CO_2Et

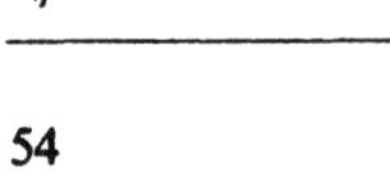

116
$t_{1/2}$ = 33 min / −27°C
P_{ij} = 0.73

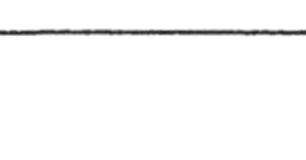

117
19 min / −27°C
0.76

the bond order of the C4-C5 bond of the thiepin is reduced. A significant contribution of *87a* to the ground state of *87* prevents the formation of a thiepin-thianorcaradiene equilibrium.

5.3 Steric Effects of Bulky Groups

As we have already mentioned in Sections 4.1, 3 and 4, it is now quite clear that the bulky groups at C-2 and C-7 of the thiepin ring play an important role in the thermal stability of the thiepin. It should again be emphasized that the 2,7-diisopropylthiepin *101* could not be detected even at —70 °C whereas the corresponding 2,7-di-*tert*-butylthiepin *105* was found to be a quite stable compound. In connection with the steric requirement of the groups substituted at C-2 and C-7 on the stability, the synthesis of the thiepin carrying an isopropyl group at C-2 and a *tert*-butyl group at C-7 such as *118* is the subject of our current attention. Furthermore, according to Eliel [56], the *tert*-butyl group ($\Delta G^0 > 4.4$ kcal mol^{-1}) may be more effective than two isopropyl groups ($\Delta G^0 = 2 \times 2.1$ kcal mol^{-1}) which suggests that the combination *tert*-butyl/methyl or even *tert*-butyl/hydrogen may be more stable than the 2,7-diisopropylthiepin *101*.

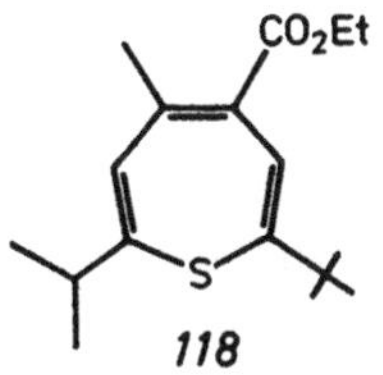

118

In 1974, Vogel and his coworkers [62] reported the first synthesis of the *syn*-benzene bisepisulfide *119* and its thermal behavior. The bisepisulfide *119* is thermolabile and decomposes at 20 °C with a half-life of about 30 min to form benzene and sulfur as final products. When the reaction was carried out in the presence of 4-phenyl-1,2,4-triazoline-3,5-dione (PTD), *119* gave the product *120* which corres-

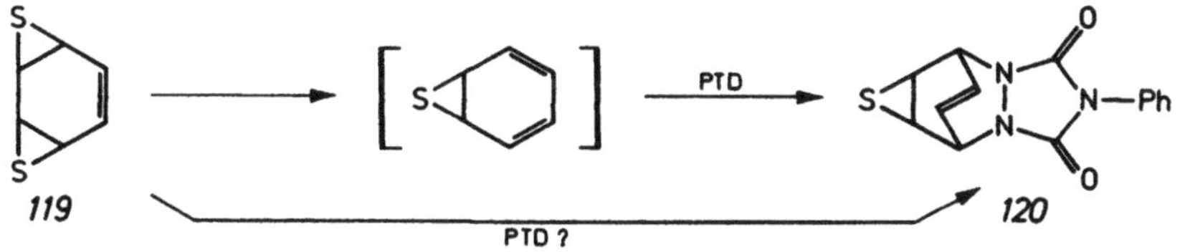

119 120

ponds to a Diels-Alder adduct of benzene episulfide (thianocaradiene) and PTD. However, it is very improbable that the benzene episulfide has sufficient lifetime to permit reaction with PTD. We feel that the adduct *120* might be formed directly from *119* prior to elimination of the first sulfur atom.

6 Mechanism of Sulfur Extrusion Reactions

One of the most important problems that has to be solved in the thiepin chemistry is the mechanism of the sulfur extrusion reaction. Ready loss of sulfur of the simplest thiepins presumably occurs by valence isomerization to the corresponding thianorcaradiene, which requires a [4n + 2] disrotatory electrocyclic process leading to a *cis*-fused three-membered ring, followed by cheletropic loss of sulfur. A lot of evidence supporting the above mechanism, though inconclusive, is available to date.

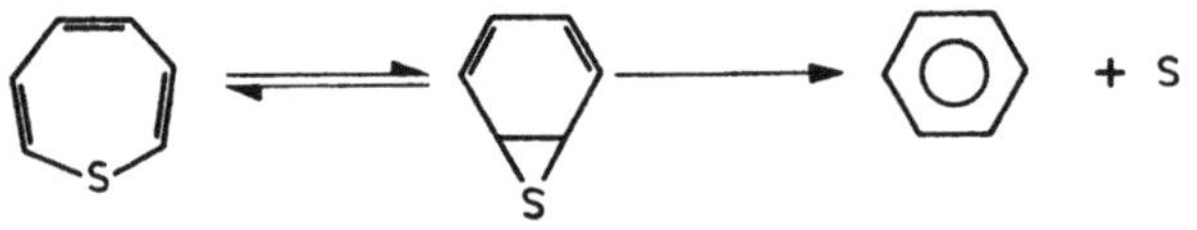

First, one of the strongest piece of evidence in support of the existence of a thianorcaradiene intermediate is the steric effect of the substituents at C-2 and C-7 of a thiepin. Substantial stability gained by 2,7-di-*tert*-butyl substitution on thiepin implies that these groups force the nonbonding interaction in the thianorcaradiene structure to be large and hence the thiepin structure will be favored (see Section 4-1, 4-3 and 4-4).

Second, substitution of the ethoxycarbonyl group at the 2-position of a thiepin ring resulted in an enhancement of the thermal stability as can be seen, for example, from a comparison between benzo[*b*]thiepin (*4*) [$t_{1/2}$ = 58 min/47 °C] and 2-ethoxycarbonylbenzo[*b*]thiepin (*73*) [$t_{1/2}$ = 352 min/47 °C). The difference in stability can be qualitatively interpreted in terms of the theory developed by R. Hoffmann and coworkers [63]. According to this theory, the orbital interaction diagram for the formation of thiirane from the interaction of an ethylene molecule with sulfur is shown in Fig. 1. The effect of substituents in an ethylene fragment is the perturbation of the π*(AS) orbital. π-Acceptors will lower the energy of the π* level implying greater interaction with the AS orbital of sulfur and hence a weaker

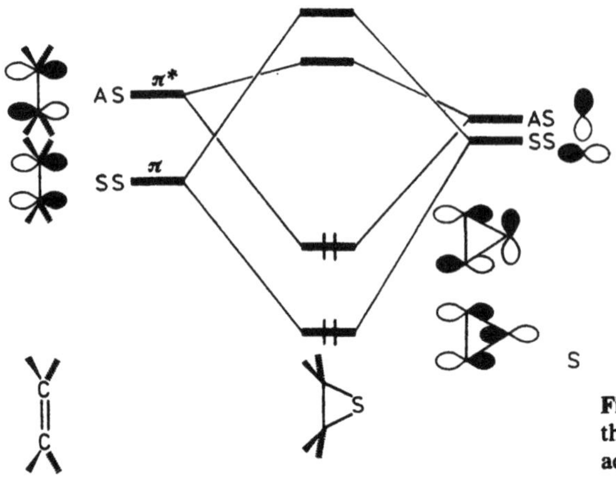

Fig. 1. Orbital interaction diagram for the formation of thiirane through interaction of ethylene and sulfur

C-C bond of the resulting thiirane. Therefore, when assuming a thianorcaradiene intermediate in the sulfur extrusion reaction of the thiepin, the relative stability of *4* and *73* can reasonably be explained.

Third, the pronounced difference in stability between the naphtho[2,3-*d*]thiepin *114* and the naphtho[1,2-*d*]thiepin *116* is a good indication of the existence of the thianorcaradiene intermediates, *114a* and *116a*, respectively in these sulfur extrusion processes. In contrast to the substantial instability of *116*, the increased stability of *114* may be due to the energy required to convert two aromatic rings into the *o*-quinoid form in *114a*.

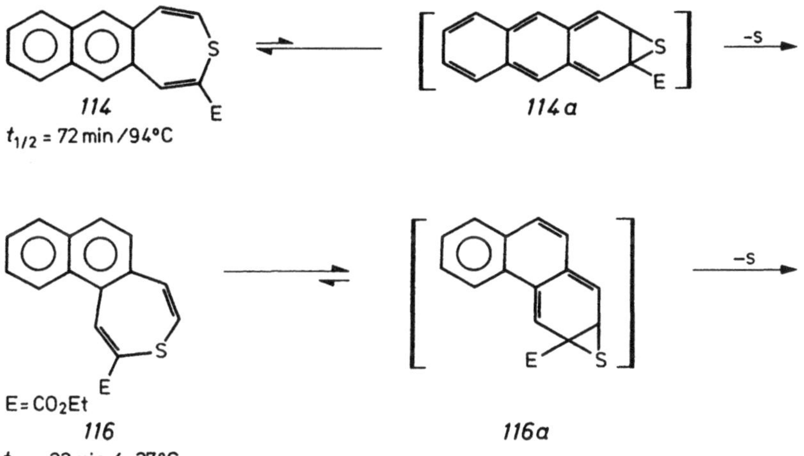

114 $t_{1/2}$ = 72 min/94°C — *114a*

E = CO_2Et

116 $t_{1/2}$ = 33 min/−27°C — *116a*

Fourth, kinetic data of the sulfur extrusion reaction of thiepin will provide direct evidence for the transition state of the process. Data on the conversion of the thiepin *34* into its corresponding naphthalene derivative are available [25]. The substantially large negative activation entropy ($\Delta S^{*} \simeq -24$ cal mol^{-1} deg^{-1}) points to the existence of a highly ordered transition state, namely a thianorcaradiene, in the reaction. 3,4-Bis(methoxycarbonyl)-5-hydroxybenzo[*b*]thiepin *33* thermally

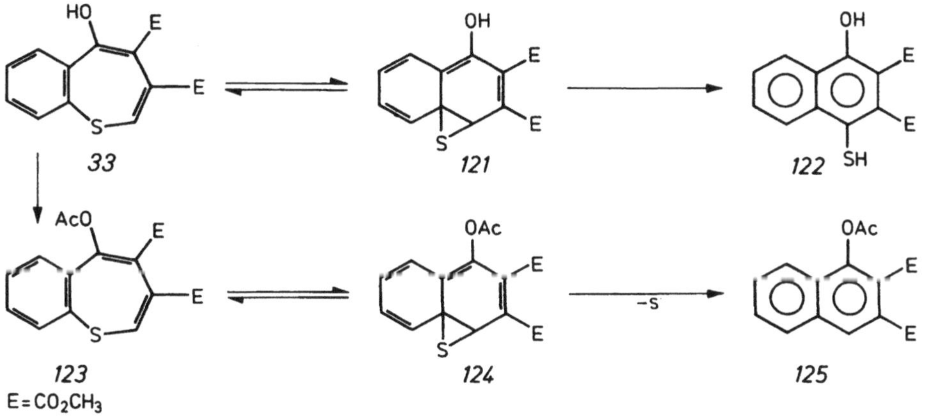

33 — *121* — *122*

123 — *124* — *125*

E = CO_2CH_3

rearranged to 2,3-bis(methoxycarbonyl)-4-mercapto-1-naphthol (*122*) with no elimination of sulfur [25]. From the activation entropy of the process ($\Delta S^{*} \simeq -25$ cal $\times mol^{-1}$ deg^{-1}), this transformation has been explained in terms of a rapid hydrogen transfer from C-2 to the sulfur atom in the intermediate thianorcaradiene (*121*). The hydrogen transfer is catalyzed by the acidic hydroxy group. This was confirmed by the acetylation of *33* (Ac_2O/AcONa, 110 °C) which yielded a sulfur-free naphthalene derivative (*125*). Obviously, the reaction proceeds via the acetate *123* and its valence isomer *124* followed by rapid loss of sulfur.

While all of these experimental facts seem to strongly suggest the intermediate existence of a thianorcaradiene, more definitive evidence might be obtained by a kinetic study of the stable monocyclic thiepin *101* for which a large negative activation entropy is to be expected. We are currently investigating this point.

7 Molecular Structure

Much attention has been focused on the molecular structure of the thiepin skeleton in order to evaluate the π-electron delocalization in 8π-heterocycles. However, there are relatively few examples of structure determination of thiepin derivatives by X-ray crystallography because of their thermal instability. The first instance is an X-ray crystallographic analysis of benzo[*b*]thiepin (*4*) at −140 °C [63]. The molecular structure projected onto the plane passing through C(2), C(3), C(6), and C(7) is shown in Fig. 2. The thiepin ring is in a boat conformation with angles of 49.1 and 30.0° between the base plane and the bow [C(2)-S-C(7)] and stern [C(3)-C(4)-C(5)-C(6)] planes, respectively. In the thiepin ring, bond alternation is clearly recognized. C(2)-C(3) and C(4)-C(5) are double bonds while the C(3)-C(4) and C(5)-C(6) bond distances are similar to those expected for $C(sp^2)$-$C(sp^2)$ single bonds.

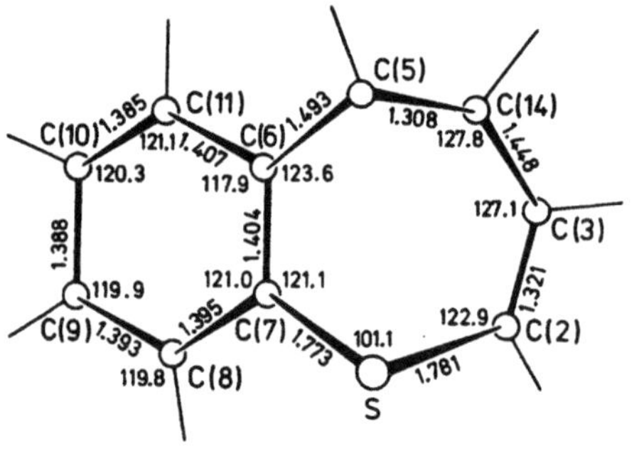

Fig. 2. Bond lengths and angles in benzo[*b*]-thiepin (*4*). Maximum errors ±0.005 Å and ±0.3°, respectively

Table 5. Angles between the base plane and the bow and stern planes of some thiepin derivatives

plane	*4*	*126*	*49*	*127*	*128*
bow	49.1	58.6	47.2	44.6	45.2
stern	30.1	40.8	24.0	22.6	19.8

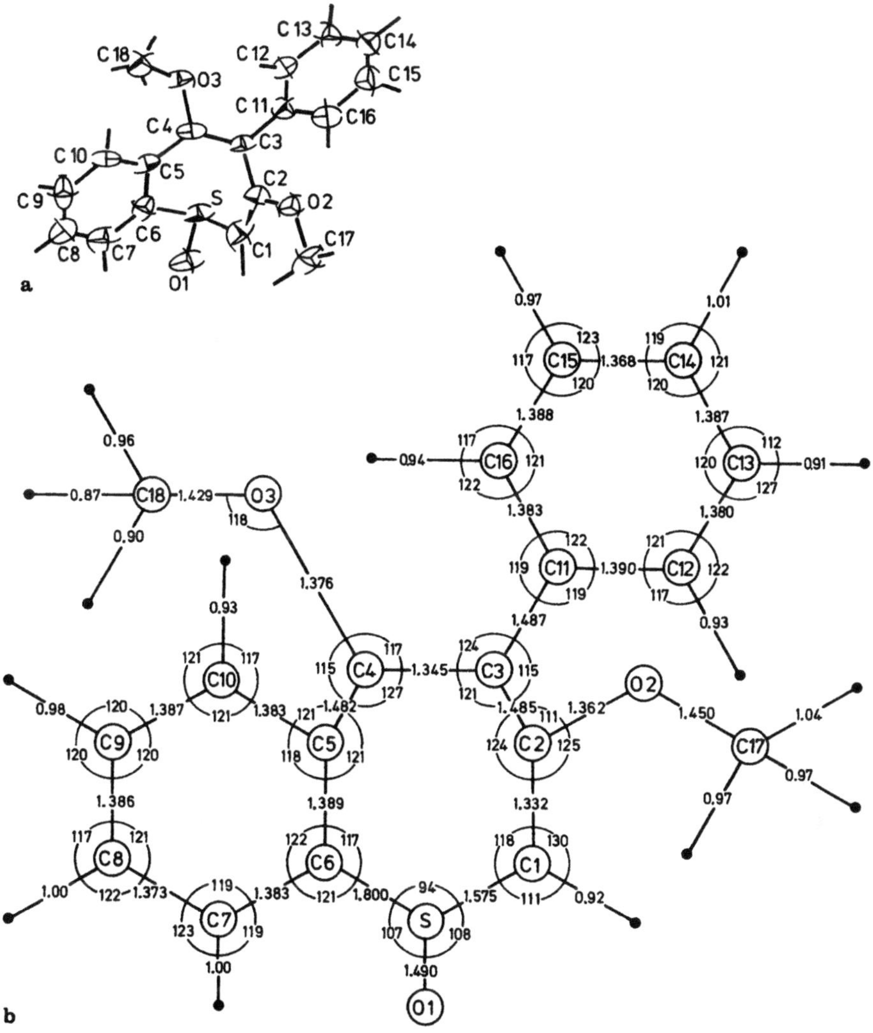

Fig. 3. Molecular structure [a] and bond lengths and angles [b] for 3,5-dimethoxy-4-phenylbenzo[*b*]-thiepin 1-oxide (*126*)

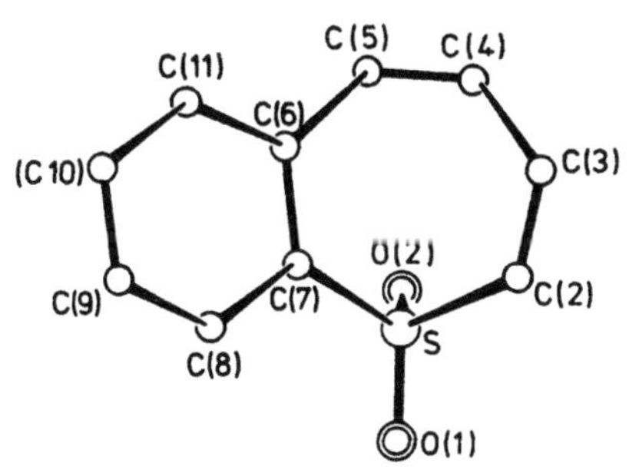

Fig. 4. Molecular structure of benzo[*b*]thiepin 1,1-oxide (*49*) projected onto the plane passing through atoms C(2), C(3), C(6), and C(7)

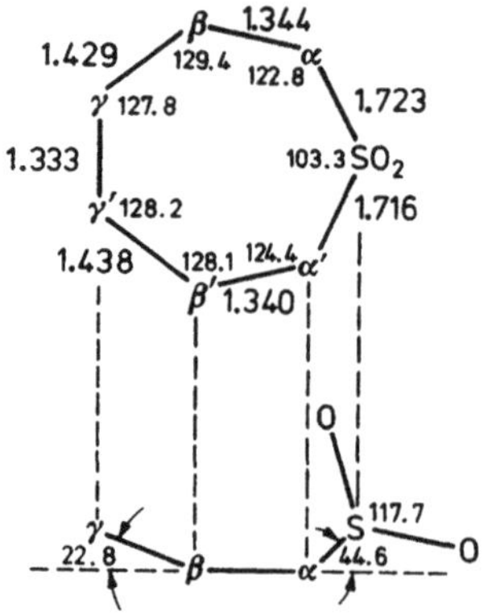

Fig. 5. Bond lengths and angles in thiepin 1,1-dioxide (*127*)

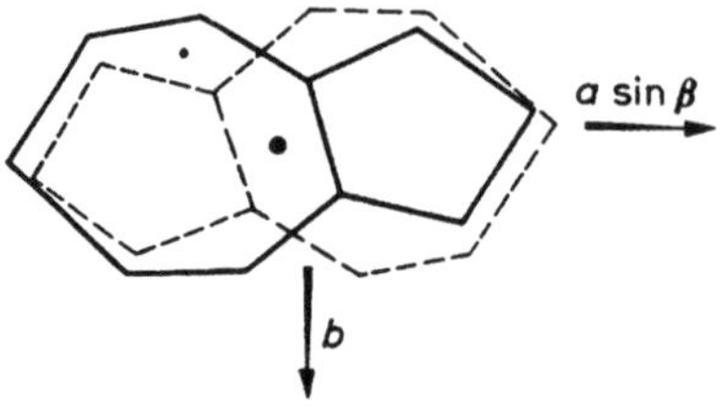

Fig. 6. Projection of a portion of the disordered thieno-[3,4-*d*]thiepin (*13*, X=S) crystal structure viewed along the c axis

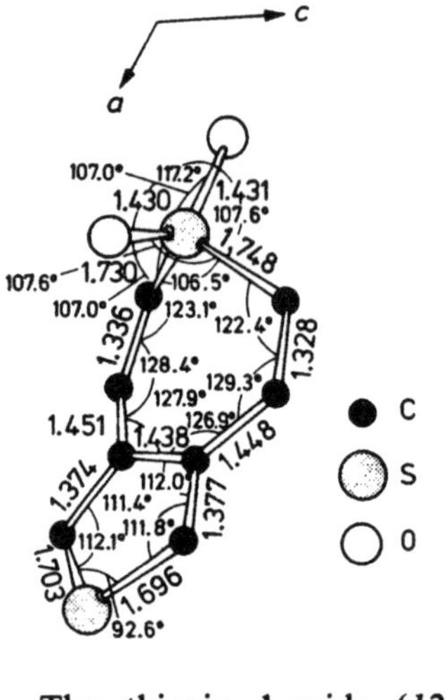

Fig. 7. Projection of thieno[3,4-*d*]thiepin 3,3-dioxide (*128*) along the b axis showing bond lengths and angles

The thiepin 1-oxide (*126*)[64] and thiepin 1,1-dioxides, such as *49*[65] and *127*[66], together with thieno[3,4-*d*]thiepin (*13*, X = S)[67] and its dioxide *128*[67] have also been analyzed by X-ray analysis. The X-ray structures are shown in Figs. 3–7.

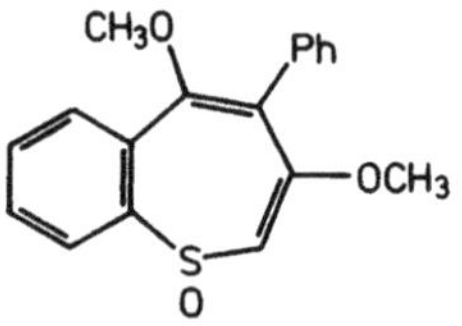

126

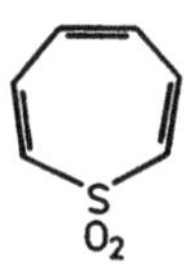

127

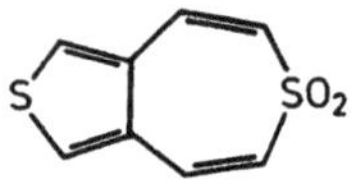

128

In all cases examined so far, the thiepin rings exist in a large valued boat conformation and the angles between the base plane and the bow and stern planes are summarized in Table 5.

8 Comments on the Antiaromaticity of Thiepin

A thiepin is formally isoelectronic with the 8π-electron 1,3,5,7-cyclooctatetraene and 1,3,5-cycloheptatrienide ion and, if planar, may actually be antiaromatic. Recently, the question of the antiaromaticity of thiepin has been the subject of interest for both synthetic and theoretical chemists. The apparent instability of the thiepin ring system is in good agreement with theoretical calculations. Dewar and Trinajstić [68] have reported that the thiepin is considered to be weakly antiaromatic (RE = —1.45 kcal mol^{-1}) based on PPP SCF MO calculations. On the other hand, Hess Jr. and Schaad [69] have found it to be substantially antiaromatic (RE = —0.232β) by using the Hückel MO method. This result was also supported by a graph-theoretical treatment by Aihara [70].

The first experimental evidence supporting the paratropicity of thiepin associated with 8π electrons has been reported by Hoffman and Schlessinger [30,71] with regard to the complex monocyclic thiepin *94*. The thiepin ring protons of *94* occur at higher field values (δ 6.50) than the corresponding protons of the furothiepin *92b* (δ 6.80), and the ortho protons of the imide phenyl group of *94* appear at lower field values (δ 6.70–7.00) than the corresponding protons of *93* (δ 6.20–6.45). From these NMR findings, they concluded that *94* might possess a paramagnetic ring current.

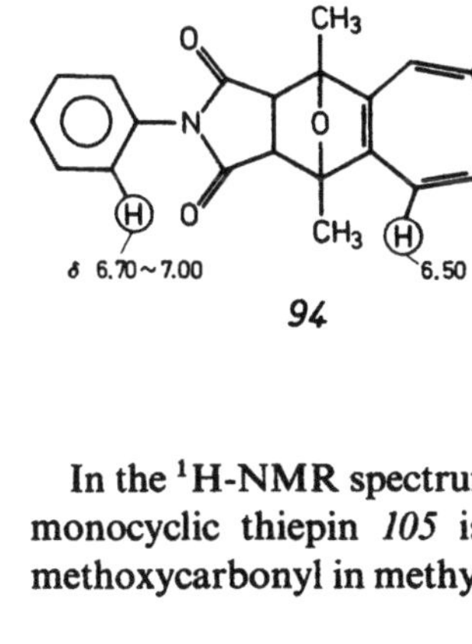

94

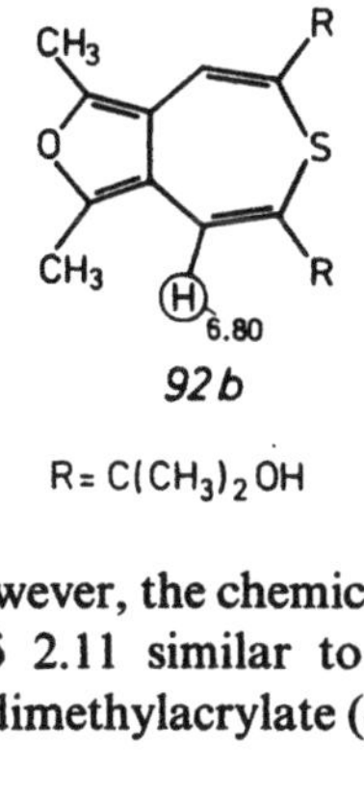

92b

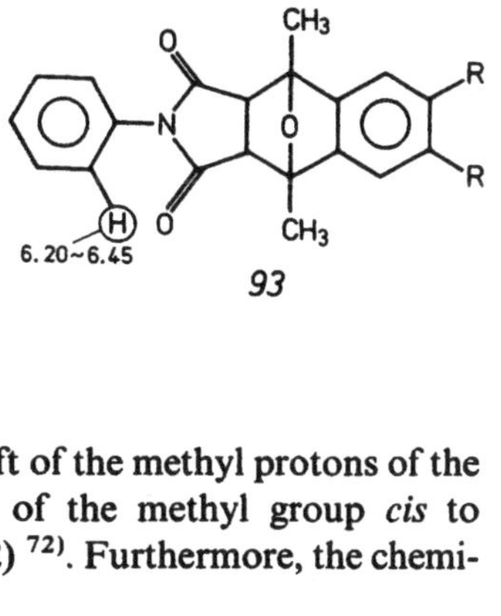

93

R = $C(CH_3)_2OH$

In the ^{1}H-NMR spectrum, however, the chemical shift of the methyl protons of the monocyclic thiepin *105* is at δ 2.11 similar to that of the methyl group *cis* to methoxycarbonyl in methyl 3,3-dimethylacrylate (δ 2.12) [72]. Furthermore, the chemi-

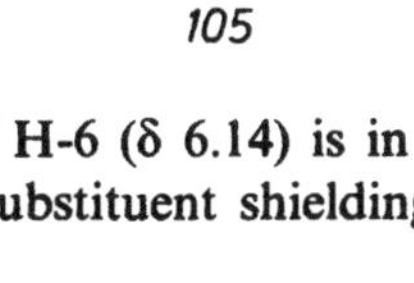

105

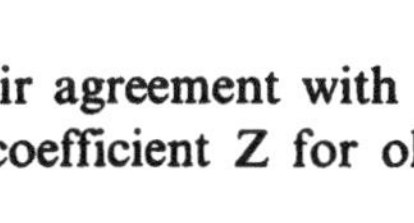

cal shift of H-6 (δ 6.14) is in fair agreement with the value (δ 6.24) calculated by using the substituent shielding coefficient Z for olefinic protons. In addition, the available X-ray results of some thiepins (see Section 7) suggest that the monocyclic thiepin *105* must exist in a boat conformation. On the basis of these results the thiepin *105* is not a paratropic but an atropic molecule. We think that the instability of thiepin may not be attributed to its antiaromaticity, but due to kinetic reasons.

The thiepin *13* (X = S) was found to be inert toward catalytic hydrogenation. The authors ascribe the stability of *13* (X = S) to a contribution of the charge-separated structures *13a* and *13b* to the ground state of *13* (X = S) [71]. An X-ray structural determination showed that *13* (X = S) is nearly planar with a disordered crystal structure similar to that of azulene [73].

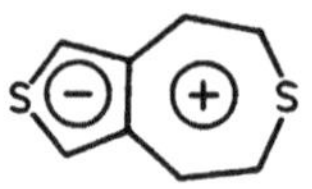

13a

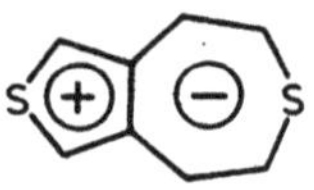

13b

CNDO/2 and extended Hückel calculations [74] of *13* (X = S) revealed a small difference in energy between the planar and nonplanar structure, both with bond alternation. These results can be translated into the valence bond structure corresponding to a cyclic thioether.

9 Thiepin 1-Oxides and Thiepin 1,1-Dioxides

The first synthesis of thiepin 1,1-dioxide (*131*) was performed by Mock in 1967 [75]. With exess bromine in chloroform 2,7-dihydrothiepin 1,1-dioxide (*129*), prepared by 1,6-addition of sulfur dioxide to *cis*-hexatriene, gave a dibromide *130* which, on treatment with two equivalents of triethylamine, afforded *131*. Upon catalytic reduction, *131* rapidly absorbed three molar equivalents of hydrogen to yield hexahydrothiepin 1,1-dioxide (*132*). The 1,1-dioxide *131* is a fairly stable compound:

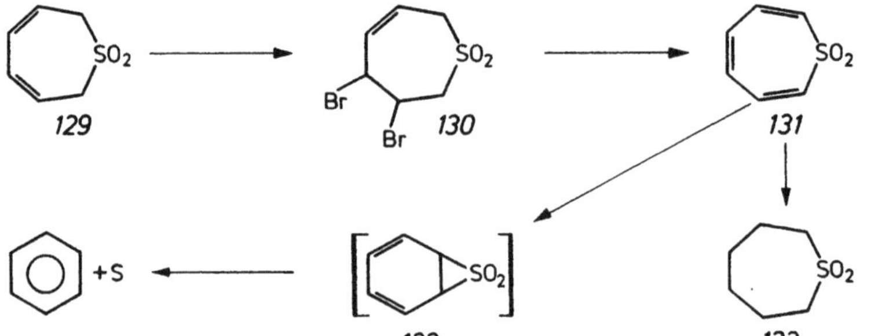

however, it decomposes at 100 °C to benzene and sulfur dioxide with a half-life of about 3 h. A probable intermediate, benzene episulfone (*133*), in this decomposition could not be detected. A priori, one may assume canonical forms such as *131a*–*131c* for the ground state of *131* (presumably involving d-orbital participation). Although the chemical and spectral properties of this molecule are consistent with a

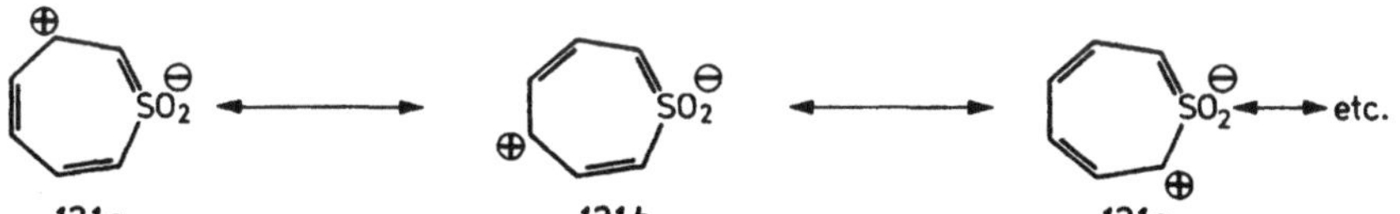

131a 131b 131c

localized triene structure, X-ray studies [66] on *131* revealed that the C(2)-S and C(7)-S bond lengths (1.723 and 1.716 Å) are abnormally short. This shortening of the C-S bonds may be due to a compromise between a minimally strained boat form and a more planar delocalized ring. The predicted double bond character of the C-S bonds in *131* has been confirmed by X-ray analysis [76]. The conformational ring inversion barrier of 3-isopropyl-6-methylthiepin 1,1-dioxide (*134*) was determined by NMR to be 6.4 kcal mol^{-1} [76]. The barrier to inversion in the parent thiepin 1,1-dioxide *131* is

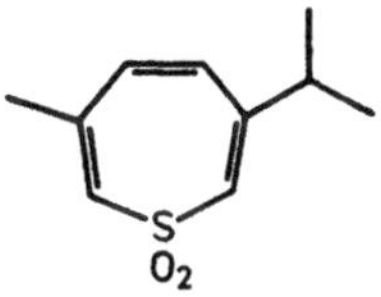

134

probably a little lower and appears to be very similar to the 6 kcal mol^{-1} barrier in 1,3,5-cycloheptatriene [76].

Several derivatives of thiepin 1,1-dioxide including the parent benzo[*b*]thiepin 1,1-dioxide have been synthesized [13,17,77,78,79,80]. In all cases, the thermal stability of these dioxides is greatly enhanced and, as a result, some of the unstable thiepins can be characterized by oxidation to their 1,1-dioxides. For example, benzo[*b*]thiepin decomposes at 47 °C with a half-life of 84 min whereas its dioxide is stable even in refluxing ethanol [81].

Unlike thiepin 1,1-dioxides, thiepin 1-oxides are very labile compounds. Treatment of the stable thieno[3,4-*d*]thiepin (*13*: X = S) with *m*-chloroperbenzoic acid in chloroform gave stable thieno[3,4-*d*]thiepin 1,1-dioxide (*135*) [54]. When the oxidation of *13* was carried out at 0 °C under carefully defined conditions, there was isolated in 90% yield the unstable crystalline thiepin 1-oxide *136* which decomposed on standing [54].

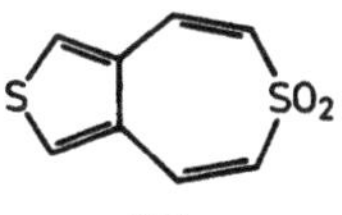

135

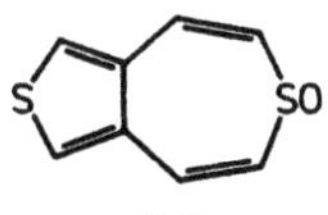

136

Systematic studies on benzo[*b*]thiepin 1-oxides were reported by Hofmann and coworkers [82,83]. 4-Phenylbenzo[*b*]thiepin 1-oxides *137a* and *137b* were prepared via the benzo[*b*] 3-thiepinone 1-oxide *138* as well as by direct oxidation of the benzo[*b*]thiepins *139a* and *139b*. The synthesis of *137c* and *137d* was only accomplished by the latter way. Further oxidation of *137* yielded the thermally stable 1,1-dioxides. Thermolysis of *137* produced the corresponding naphthalenes by

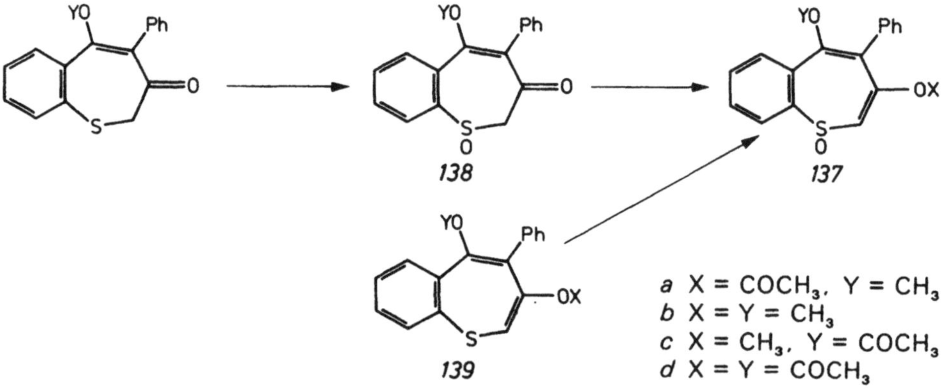

138 137 139

a X = $COCH_3$, Y = CH_3
b X = Y = CH_3
c X = CH_3, Y = $COCH_3$
d X = Y = $COCH_3$

elimination of the sulfinyl group. The reaction rates and the half-lives of the SO elimination reactions measured by NMR spectroscopy are summarized in Table 6 together with those of the thiepins *139*. It is apparent that the thiepin 1-oxides are less stable than the corresponding thiepins and, analogously to benzo[*b*]thiepin, the thermal stability of the sulfoxide ring is increased by electron-donating groups. This trend is in good agreement with our observation on benzo[*b*]thiepins.

In connection with thiepin 1,1-oxides, 1-methylbenzo[*b*]thiepinium salts *140* have been synthesized [84]. These salts are thermally stable and the half-life of *140a*

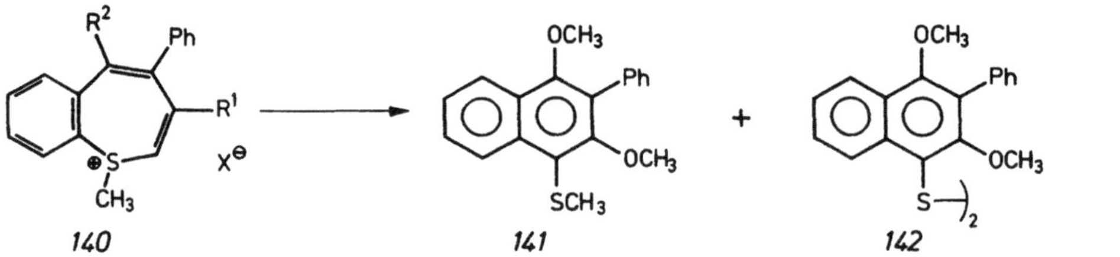

140 141 142

a $R^1 = R^2 = OCH_3$, $X = FSO_3$ *b* $R^1 = OCOCH_3$, $R^2 = OCH_3$, $X = FSO_3$ *c* $R^1 = R^2 = OCOCH_3$, $X = SO_2C_6H_2(NO_2)_3$

was found to be 69 h at 80 °C in CD_3NO_2. The thermolysis of *140a* in dimethylformamide gave the naphthalenes *141* and *142* whereas when the thermolysis was carried out in dimethyl sulfoxide, *140a* was converted to the corresponding benzo[*b*]thiepin via rapid transmethylation.

Table 6. Rate constants and half-lives of the SO elimination reactions of *137* and *139*

137	k [s^{-1}] $\times 10^{-4}$	$t_{1/2}$ [min]	*139*	k [s^{-1}] $\times 10^{-4}$	$t_{1/2}$ [min]
a	2.2/54 °C	52	*a*	1.75/89 °C	66
b	2.5/50 °C	45	*b*	2.5 /90 °C	45
c	2.5/20 °C	45	*c*	0.64/59 °C	180
d	2.8/15 °C	41	*d*	2.75/64 °C	42

10 Dithiepin Systems

Since divalent sulfur is isoelectronic with the carbon-carbon double bond and can participate in π-electron delocalization, the anions derived from 1,2- (*143*), 1,3- (*144*), and 1,4-dithiepins (*145*) may be regarded as 10π-electron aromatic systems. According to the prediction by Zahradnik and Parkanyi [85] based on Hückel's molecular

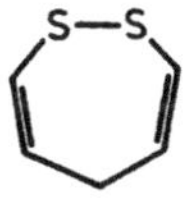

143 *143a* *144* *144a* *145* *145a*

orbital calculations, all three anions (*143a*, *144a* and *145a*) should be unstable and difficult to prepare. A pioneering study on *144a* was reported by Breslow and Mohacsi [86] in 1963. From a detailed examination of the deuterium exchange behavior of ethyl dibenzo[*b*, *d*] [1, 3]dithiepin-6-carboxylate (*146*) and the comparison of its properties with those of the open-chain analog *147*, they concluded that no aromaticity associated with the cyclic 10π-electron system could be observed. On the other hand, the parent 1,3-dithiepin anion (*144a*) has been synthesized and

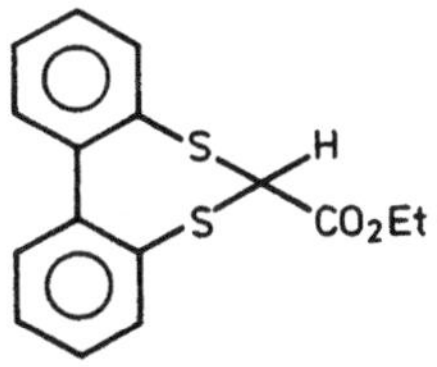
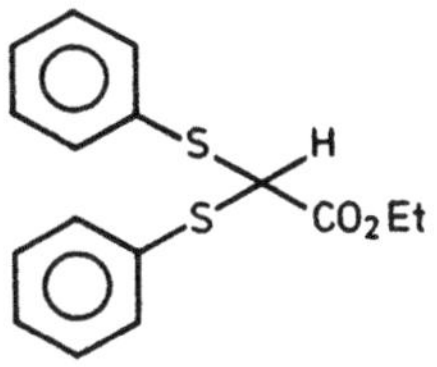

146 *147*

shown by Semmelhack, Chiu and Grohmann [87] to be the potential 10π-aromatic system. The synthesis was started from 1,3-dithia-5-cycloheptene (*148*) [88] via a sequence of reactions shown below. The reaction of *144* with *n*-butyllithium in tetra-

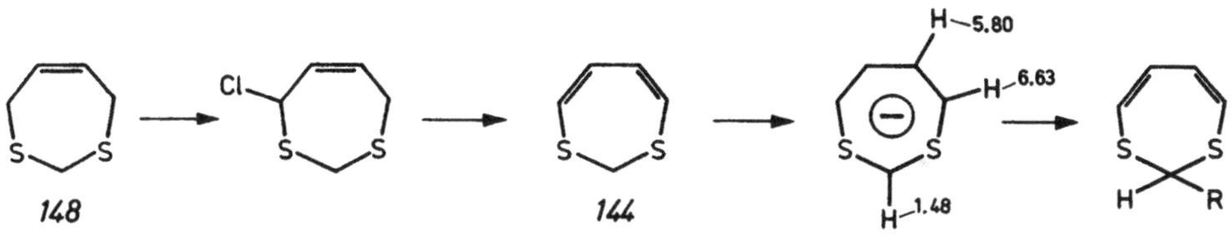

148 *144* *144a* *149*

hydrofuran under argon at —40 °C yields a deep red solution of the anion *144a* which was quenched with electrophilic reagents to give *149*. The chemical shifts indicated in *144a* suggested a certain delocalization of the negative charge. The unusually high field absorption of the proton at position 2 was proposed to be due to the involvement of sulfur d-orbitals. The small W-letter coupling of 0.8 Hz observed between H-2 and H-4, 7 was attributed to a relatively planar structure of *144a*. These results seem to suggest that the anion *144a* represents a 10π-heteroaromatic system.

Shortly after, the same authors reported the results on the relative acidities of *144* as compared with 1,3-dithiacycloheptane (*150*) and the open-chain model bis-(vinylthio)methane (*151*)[89]. In refluxing *tert*-butyl alcohol-*d* in the presence of

150

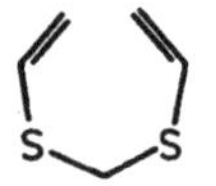

151

potassium *tert*-butoxide, *144* completely exchanges both protons for deuterium in position 2 within 10 min. Under the same conditions, the saturated model *150* does not exhibit H/D exchange at all. Furthermore, *144* has nearly completely exchanged its protons at C-2 in the same solvent system after 30 min at 50 °C while the open-chain analog *151* does hardly undergo any H/D exchange reaction. From these results, the authors concluded that the 1,3-dithiepin anion *144a* represents a 10π-heteroaromatic system, although the degree of aromaticity seems to be small.

Some fulvalene-type compounds containing 1,3-dithiepin counterparts such as *152*, *153* and *154* have been synthesized by reaction of 2-trimethylsilyl-1,3-dithiepin anion with di-*tert*-butylcyclopropenone, 2,3; 6,7-dibenzotropone, and 9-fluorenone, respectively[90]. The most characteristic feature of these fulvalenes can be seen in the ^{1}H-NMR signals of the four vinyl protons attached to the 1,3-dithiepin moiety. These protons in 2*H*-dithiepin (*144*) show the very narrow AA'BB' signals centered at δ 6.31 whereas the same protons in the anion *144a* reveal a well separated AA'XX' pattern[87]. The substantial upfield shift of H-5,6 of *144a* compared to those of *144* has been attributed to the considerable charge density at these positions in *144a*. Therefore, it can be assumed that the degree of the chemical shift difference between H-4,7 and H-5,6 reflects the anionic character of the 1,3-dithiepin moieties in the fulvalenes. The observed chemical shifts of the vinyl protons of *152*, *153* and *154* are summarized in Table 7. While the fulvalene *152* having a three-membered ring shows moderately separated AA'BB' signals, *154* exhibits an extremely narrow doublet accompanied by small satellites. It is clear

Table 7. Chemical shifts of vinyl protons in *152–154*

	 *144a*		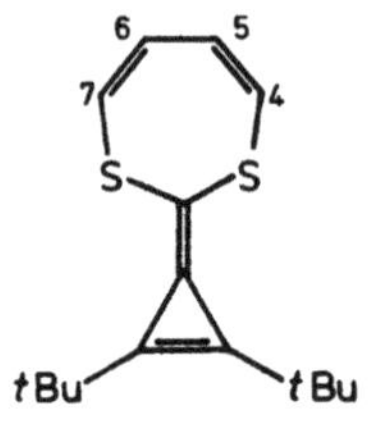 *152*		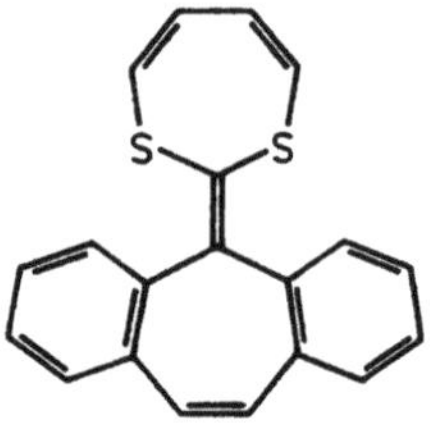 *153*		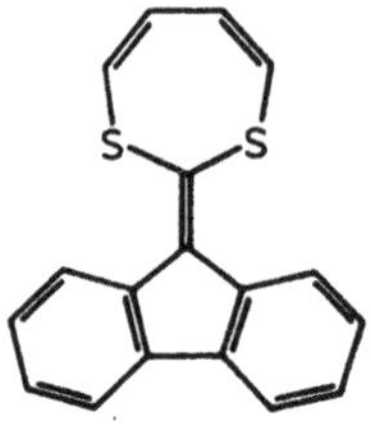 *154*		 *144*
$\delta_{5,6}$:	5.80		5.96		6.14		6.52		6.29
$\delta_{4,7}$:	6.63		6.32		6.27		6.57		6.33
$\Delta\delta$:	0.83	>	0.36	>	0.13	>	0.05	~	0.04

that the chemical shift difference, $\Delta\delta = \delta_{4,7} - \delta_{5,6}$, increases in the order *144a* > *152* > *153* > *154* ~ *144*, parallel to the electron-donating ability of the carbocyclic counterparts. Thus, the 1,3-dithiepin ring in *152* will tend to stabilize the dipolar form by assiting delocalization of the negative charge. The same is true, to a certain extent, of *153*.

Unlike 1,3-dithiepin anion *144a*, the evidence for the instability of *145a* and for the lack of aromaticity associated with 10π-electron delocalization through the sulfur atom has been reported[91,92]. The reaction of the disodium salt of *cis*-dimercaptoethylene (*155*) with either 1,2-dibromo-3-propanol or 1,3-dibromo-2-propanol yielded 6,7-dihydro-5*H*-1,4-dithiepin-6-ol (*156*). Treatment of the methoxy derivative *157* derived from *156* with two equivalents of lithium dicyclohexylamide resulted in an effective elimination of methanol to give 5*H*-1,4-dithiepin (*145*) as a colorless liquid. Lithiation of *145* with *n*-butyllithium in tetrahydrofuran at —70 °C

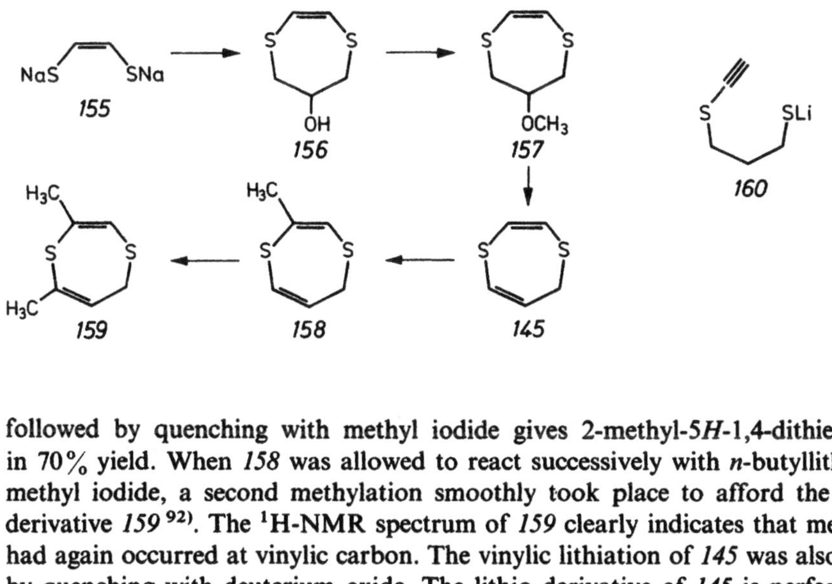

followed by quenching with methyl iodide gives 2-methyl-5*H*-1,4-dithiepin (*158*) in 70% yield. When *158* was allowed to react successively with *n*-butyllithium and methyl iodide, a second methylation smoothly took place to afford the dimethyl derivative *159*[92]. The ^{1}H-NMR spectrum of *159* clearly indicates that methylation had again occurred at vinylic carbon. The vinylic lithiation of *145* was also revealed by quenching with deuterium oxide. The lithio derivative of *145* is perfectly stable at —40 °C but undergoes ring opening at —15 °C to the corresponding 1-alkyne *160*[91].

11 Conclusion

In this article, a description has been given of recent investigations in the field of thiepin chemistry which are important for the major seven-membered unsaturated heterocyclic compounds. While the properties of these sulfur heterocycles have been determined quite well, some important questions still remain unexplored. Our current research efforts involve attempts to understand more generally the substituent effects on the thermal stability of the thiepin.

The research described in this article has been financially supported in part by a Grant-in-Aid for Scientific Research (No. 343007) from the Ministry of Education, Japan. We are grateful to the many co-workers whose names appear in the literature compilation. Dr. Thomas L. Tarnowski, Syva Research Institute, kindly read the manuscript, and, in addition, gave valuable scientific comments. Finally, one of the authors (I. M.) would like to express his sincere thanks to Professor Emeritus Tetsuo Nozoe for his encouragement.

Note added in Proof

An alternative synthesis of thiepin 1,1-dioxide (*131*) has also been reported (Paquette, L. A., Maiorana, S.: J. Chem. Soc. Chem. Commun. 1971, 313). Thus, the reaction of vinyldiazomethane with sulfur dioxide yields 4,5-dihydrothiepin 1,1-dioxide which was converted to *131* by bromination (NBS) and dehydrobromination (Et_3N) sequence.

12 References

1. For a review see Paquette, L. A., in: Nonbenzenoid Aromatics (ed.) Snyder, J. P., Vol. 1, p. 249 Academic Press, New York 1969; Paquette, L. A.: Angew. Chem. Int. Ed. Engl. *10*, 11 (1971)
2. For reviews see Vogel, E., Günther, H.: Angew. Chem. Int. Ed. Engl. *6*, 385 (1967); Rosowsky, A., in: Seven-Membered Heterocyclic Compounds Containing Oxygen and Sulfur (ed.) Rosowsky, A., Chapts. I–IV, Wiley-Interscience, New York 1972; Jerina, D. M., Yagi, H., Daly, J. W.: Heterocycles *1*, 267 (1973)
3. For an attempted synthesis of parent thiepin see Barton, T. J., Martz, M. D., Zika, R. G.: J. Org. Chem. *37*, 552 (1972)
4. Field, L., Tuleen, D. L., Traynelis, V. J., in: Seven-Membered Heterocyclic Compounds Containing Oxygen and Sulfur (ed.) Rosowsky, A., Chapts. X, XI, Wiley-Interscience, New York 1972; Eisner, U., Krishnamurthy, Int. J. Sulfur Chem. *6*, 267 (1971); Reid, D. H.: Org. Compd. Sulfur, Selenium, Tellurium *1*, 358 (1970); *2*, 545 (1973); Eisner, U.: Org. Compd. Sulfur, Selenium, Tellurium *4*, 332 (1977)
5. Traynelis, V. J., Livingston, Jr., J. R.: J. Org. Chem. *29*, 1092 (1964)
6. Traynelis, V. J. et al.: J. Org. Chem. *38*, 3978 (1973)
7. Jilek, J. O. et al.: Collect. Czech. Chem. Commun. *32*, 3186 (1967)
8. Jilek, J. O. et al.: Monatsh. Chem. *96*, 182 (1965)
9. Fouche, J.: Bull. Soc. Chim. France, *1970*, 1376
10. Chem. Abstr. *65*, 12183 (1966)
11. Schlessinger, R. H., Ponticello, G. S.: J. Am. Chem. Soc. *89*, 7138 (1967)
12. Schlessinger, R. H., Ponticello, G. S.: Tetrahedron Lett. *1969*, 4361
13. Hofmann, H., Westernacher, H.: Angew. Chem. Int. Ed. Engl. *5*, 958 (1966); Chem. Ber. *102*, 205 (1969)
14. Hofmann, H., Meyer, B., Hofmann, P.: Angew. Chem. Int. Ed. Engl. *11*, 423 (1972); Hofmann, H. et al.: Chem. Ber. *108*, 3596 (1975)
15. Bergmann, E. D., Rabinovitz, M.: Israel J. Chem. *1*, 125 (1963)
16. Urberg, M. M., Kaiser, E. T.: J. Am. Chem. Soc. *89*, 5931 (1967)
17. Bergmann, E. D., Rabinovitz, M.: J. Org. Chem. *25*, 828 (1960)
18. Whitlock, H. W.: Tetrahedron Lett. *1961*, 593
19. Seidl, H., Biemann, K.: J. Heterocycl. Chem. *4*, 209 (1967)
20. Schweizer, E. E., Parham, W. E.: J. Am. Chem. Soc. *82*, 4085 (1960)
21. Parham, W. E., Weetman, D. G.: J. Org. Chem. *34*, 56 (1969)

22. Traynelis, V. J. et al.: J. Org. Chem. *43*, 3379 (1978)
23. Neckers, D., Dopper, J., Wynberg, H.: Tetrahedron Lett. *1969*, 2913
24. Reinhoudt, D. N., Kowenhoven, C. G.: J. Chem. Soc. Chem. Commun. *1972*, 1232
25. Reinhoudt, D. N., Kowenhoven, C. G.: Tetrahedron *30*, 2431 (1974)
26. Scott, G. P.: J. Am. Chem. Soc. *75*, 6332 (1953)
27. Dimroth, K., Lenke, G.: Angew. Chem. *68*, 519 (1956)
28. Dallacker, F., Glombitza, K. W., Lipp, M.: Justus Liebigs Ann. Chem. *643*, 82 (1961)
29. Loudon, J. D., Sloan, A. B. D.: J. Chem. Soc. *1962*, 3262
30. Hoffman, Jr. J., Schlessinger, R. H.: J. Am. Chem. Soc. *92*, 5263 (1972)
31. Büschken, W.: Dissertation, Köln 1972
32. Cf. Vogel, E.: Israel J. Chem. *20*, 215 (1980)
33. Loudon, J. D., Sloan, A. B. D., Summers, L. A.: J. Chem. Soc. *1957*, 3814
34. Neale, A. J., Rawlings, T. J., McCall, E. B.: Tetrahedron *21*, 1299 (1965)
35. For a review see Cardin, D. J. et al.: Chem. Soc. Rev. *2*, 99 (1973)
36. Murata, I., Nakasuji, K.: Tetrahedron Lett. *1973*, 47; Pagni, R. M., Watson, C. R.: ibid. *1973*, 59
37. Katz, T. J., Wang, E. J., Acton, N.: J. Am. Chem. Soc. *93*, 3782 (1971)
38. Murata, I., Tatsuoka, T., Sugihara, Y.: Tetrahedron Lett. *1973*, 4261
39. See e.g. Jackman, L. M., Sternhell, S.: Application of Nuclear Magnetic Resonance Spectroscopy in Organic Chemistry, 2nd edit., p. 335, Pergamon Press, Oxford 1969
40. Kabuto, C. et al.: Angew. Chem. *86*, 738 (1974); Angew. Chem. Int. Ed. Engl. *13*, 669 (1974)
41. Murata, I., Tatsuoka, T., Sugihara, Y.: Angew. Chem. *86*, 161 (1974); Angew. Chem. Int. Ed. Engl. *13*, 142 (1974)
42. Murata, I., Tatsuoka, T., Sugihara, Y.: Tetrahedron Lett. *1974*, 199
43. Tatsuoka, T.: Dissertation, Osaka University 1975
44. Murata, I., Tatsuoka, T.: Tetrahedron Lett. *1975*, 2697
45. Masamune, S. et al.: J. Am. Chem. Soc. *95*, 8481 (1973)
46. Lüttringhaus, A., Kolb, A.: Z. Naturforsch. *16b*, 762 (1961)
47. Schöllkopf, U., Frasnelli, H.: Angew. Chem. *82*, 291 (1970); Angew. Chem. Int. Ed. Engl. *9*, 301 (1970)
48. Nakasuji, K. et al.: Angew. Chem. *88*, 650 (1976); Angew. Chem. Int. Ed. Engl. *15*, 611 (1976)
49. Ishihara, T.: PhD Thesis, Osaka University 1980
50. Nishino, K., Nakasuji, K., Murata, I.: Tetrahedron Lett. *1978*, 3567
51. Wynberg, H., Helder, R.: Tetrahedron Lett. *1972*, 3647
52. Reinhoudt, D. N. et al.: Tetrahedron Lett. *1972*, 5269
53. a) Reinhoudt, D. N., Kouwenhoven, C. G.: J. Chem. Soc. Chem. Commun. *1972*, 1233; b) Reinhoudt, D. N., Kouwenhoven, C. G.: Tetrahedron *30*, 2093 (1974)
54. Schlessinger, R. H., Ponticello, G. S.: Tetrahedron Lett. *1968*, 3017
55. Yano, S. et al.: Chem. Lett. *1978*, 723
56. A popular measure of steric requirements is *A*, the free energy difference involved in the axial-equatorial equilibrium of substituted cyclohexane. Recommended *A*-values for *i*-C_3H_9 and tert-C_4H_9 are 2.1 and >4.4, respectively. Eliel, E. L.: Chem. Educ. *37*, 126 (1960); Angew. Chem. Int. Ed. Engl. *4*, 761 (1965)
57. Nishino, K. et al.: J. Am. Chem. Soc. *101*, 5059 (1979)
58. Kohashi, Y., Yamamoto, K., Murata, I.: unpublished results
59. Hess, Jr., B. A., Schaad, L. J., Reinhoudt, D. N.: Tetrahedron *33*, 2683 (1977)
60. Inagaki, S., Hirabayashi, Y.: J. Am. Chem. Soc. *99*, 7418 (1977)
61. The same tendency has also been observed in the annelated 1,2-diazepines, see Tsuchiya, T., Enkaku, M., Sawanishi, H.: Heterocycles *9*, 621 (1978)
62. Vogel, E., Schmidbauer, E., Altenbach, H.-J.: Angew. Chem. *86*, 818 (1974); Angew. Chem. Int. Ed. Engl. *13*, 736 (1974)
63. Yasuoka, N. et al.: Angew. Chem. *88*, 295 (1976); Angew. Chem. Int. Ed. Engl. *15*, 297 (1976)
64. Hofmann, H., Bohme, R., Wilhelm, E.: Chem. Ber. *111*, 309 (1978)
65. Yasuoka, N., Kai, Y., Kasai, N.: Acta Crystallogr. *B31*, 2729 (1975)
66. Ammon, H. L. et al.: J. Am. Chem. Soc. *90*, 4501 (1968)
67. Sakore, T. D., Schlessinger, R. H., Sobell, H. M.: J. Am. Chem. Soc. *91*, 3995 (1969)
68. Dewar, M. J. S., Trinajstić, N.: J. Am. Chem. Soc. *92*, 1453 (1970)
69. Hess, Jr. B. A., Shaad, L. J.: J. Am. Chem. Soc. *95*, 3907 (1973)

70. Aihara, J.: J. Am. Chem. Soc. *98*, 2750 (1976)
71. Schlessinger, R. H.: Aromaticity, pseudo-aromaticity, anti-aromaticity, in: The Jerusalem Symposia on Quantum Chemistry and Biochemistry (eds.) Bergmann, E. D., Pullman, B., p. 158, The Israel Academy of Sciences and Humanities, 1971
72. Jackman, L. M., Wiley, R. H.: J. Chem. Soc. *1960*, 2881
73. Sakore, T. D., Schlessinger, R. H., Sobell, H. M.: J. Am. Chem. Soc. *91*, 3995 (1969)
74. Gleiter, R. et al.: J. Am. Chem. Soc. *95*, 2860 (1973)
75. Mock, W. L.: J. Am. Chem. Soc. *89*, 1821 (1967)
Ammon, H. L., Watts, Jr. P. H., Stewart, J. M.: Acta Crystallogr. *B26*, 1079 (1970)
76. Anet, F. A. L. et al.: J. Am. Chem. Soc. *91*, 7782 (1969)
77. Hofmann, H., Westernacher, H.: Chem. Ber. *102*, 205 (1969)
78. Hofmann, H., Westernacher, H.: Angew. Chem. *79*, 238 (1967); Angew. Chem. Int. Ed. Engl. *6*, 255 (1967)
79. Hofmann, H., Westernacher, H., Haberstroh, H. J.: Chem. Ber. *102*, 2595 (1969)
80. Truce, W. E., Lotspeich, F. J.: J. Am. Chem. Soc. *78*, 848 (1956)
81. Traynelis, V. J., Love, R. F.: J. Org. Chem. *29*, 366 (1964)
82. Hofmann, H., Gaube, H.: Angew. Chem. *87*, 843 (1975); Angew. Chem. Int. Ed. Engl. *14*, 812 (1975)
83. Hofmann, H., Gaube, H.: Chem. Ber. *112*, 781 (1979)
84. Hofmann, H., Molnar, A.: Tetrahedron Lett. *1977*, 1985
85. Zahradnik, R., Parkanyi, C.: Collect. Czech. Chem. Commun. *30*, 3016 (1965)
86. Breslow, R., Mohacsi, E.: J. Am. Chem. Soc. *85*, 431 (1963)
87. Semmelhack, C. L., Chiu, I.-C., Grohmann, K. G.: Tetrahedron Lett. *1976*, 1251
88. Friebolin, H. et al.: Tetrahedron Lett. *1964*, 1929
89. Semmelhack, C. L., Chiu, I.-C., Grohmann, K. G.: J. Am. Chem. Soc. *98*, 2005 (1976)
90. Sugihara, Y., Fujiyama, Y., Murata, I.: Chem. Lett. *1980*, 1427
91. Chiu, I.-C., Grohmann, K. G.: Abstr. Papers, 169th Nat. Meet. Am. Chem. Soc., ORGN 94, 1975
92. Murata, I., Nakasuji, K.: Tetrahedron Lett. *1975*, 1895

Short-Lived Phosphorus(V) Compounds Having Coordination Number 3

Manfred Regitz and Gerhard Maas

Fachbereich Chemie der Universität D-6750 Kaiserslautern, Germany

Table of Contents

1 Introduction

In the course of the tempestuous development of organophosphorus chemistry, interest has only recently been focused on compounds of formally quinquevalent phosphorus having coordination number 3, such as *1*, *2*, or *3*, although one of the other species of this kind has long been postulated as reactive intermediate of solvolysis of phosphorylation reactions. Definite evidence of even proof of the existence of such coordinatively unsaturated phosphorus compounds, however, has been obtained only recently in mechanistic studies, by trapping reactions with suitable cycloaddends, or actually by direct isolation.

Thus it has so far proved possible to isolate stable derivatives of monomeric metaphosphoric acid and of metathio- and metaselenophosphoric acid, which, understandably, generally bear tert-butyl and/or trimethylsilyl substituents [1]. Specifically, we know aminobisiminophosphoranes (*3*, Z = NR_2, X = Y = NR) [2,3,4], aminoiminothio (or seleno)phosphoranes (*3*, Z = NR_2, X = NR, Y = S or Se) [5], and aminoiminomethylenephosphoranes (*1*, R = NR_2, X = NR) [6]. Conspicuously, no stable phosphorus(V) three-coordinate compounds have been synthesized with oxygen as divalent ligand.

Compounds of types *1*, *2*, and *3* were deduced to have planar, trigonal configurations from kinetic and stereochemical observations [7]; however, confirmation of this postulate was provided only a few years ago by the X-ray structure analysis of (bistrimethylsilyl)aminobis(trimethylsilylimino)phosphorane [8]. The P/N(imine) bonds were found to be relatively short, which was attributed to a high π-bonding component in the planar system of coordinatively unsaturated phosphorus.

metaphosphinates	metaphosphonates	metaphosphates
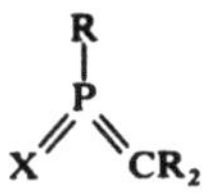		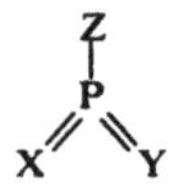
X=O, S	X=O, NR	Z=$O^{\ominus}$, OR, NR_2 X, Y=O, S, NR_2
1	*2*	*3*

The present survey is concerned exclusively with short-lived compounds of quinquevalent phosphorus with coordination number 3, which are not yet isolable in a classical sense; with few exceptions they all possess at least one P/O double bond. Specifically, these are the metaphosphinates *1*, the metaphosphonates *2*, and the metaphosphates *3*. Studies of the methyleneoxophosphoranes *1* (X = O) and the monomeric metaphosphate ion *3* (Z = $O^{\ominus}$, X = Y = O) have been especially thorough.

2 Metaphosphinates

Phosphoryl-substituted diazo compounds of general type *4* have recently been synthesized by amine diazotization, Bamford-Stevens reaction, and diazo group transfer [9]. They represent an excellent source for the generation and study of the reactivity of metaphosphinates (*6*). Photolysis and thermolysis of the diazo compounds *4* initially afford the phosphorylated carbenes *5* [10] which — in addition to other reactions — are also capable of 1,2-R shifts from phosphorus to the carbene carbon atom, leading to the metaphosphinates *6*.

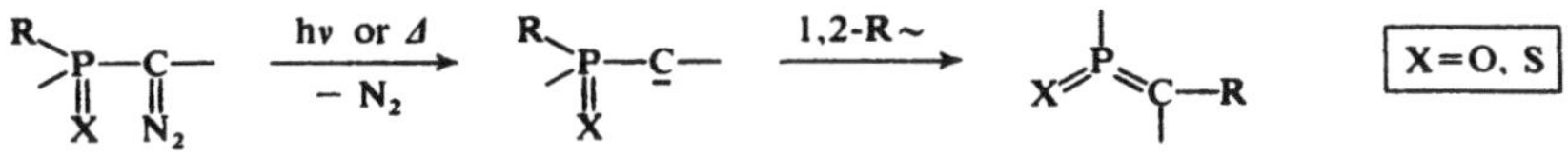

4 *5* *6*

The overall reaction is reminiscent of the Wolff rearrangement of α-diazo carbonyl compounds which gives ketenes. Compounds *6* formally represent their phosphorus analogues and are sometimes also designated as phosphenes [11].

2.1 Arylmethyleneoxophosphoranes

As a proven technique for detecting reactive intermediates, flash pyrolysis seems to be the method of choice for direct detection or isolation of phosphenes. The result of thermal decomposition of (α-diazobenzyl)diphenylphosphine oxide (*7*) was nevertheless disappointing, since only triphenylmethane (*13*), fluorene (*14*), and benzophenone (*15*) but not the desired methyleneoxophosphorane *9* could be isolated [12].

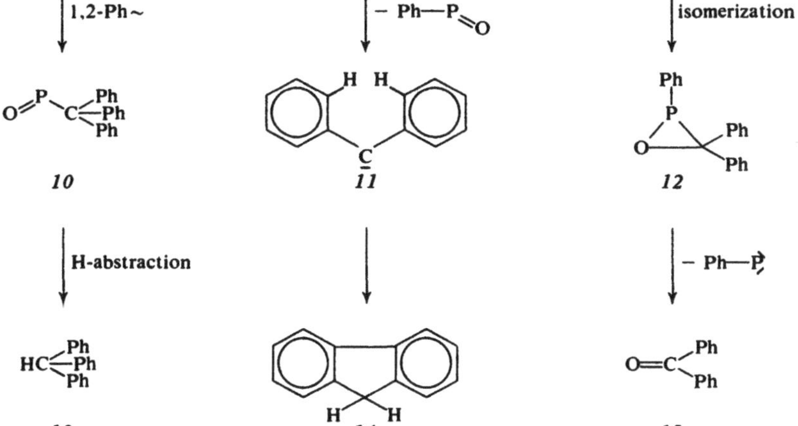

7 *8* *9* *10* *11* *12* *13* *14* *15*

Presumably, *9* is actually formed from carbene *8* in the pyrolysis zone by a P/C phenyl shift, but then apparently succumbs to fast transformation into the thermodynamically stable final products. Formation of the methane derivative *13* should be preceded by a 1,2-phenyl shift to give the shortlived *10*, the production of fluorene (*14*) by the occurrence of diphenylcarbene (*11*), and the formation of benzophenone (*15*) by isomerization to the angle-strained three-membered heterocycle *12*, which is followed by elimination of phenylphosphinidene. No direct evidence is available for the intermediacy of *10–12*.

In contrast to the situation on flash pyrolysis, methyleneoxophosphoranes generated by thermolysis or photolysis in the presence of protic nucleophiles can be directly trapped to form corresponding derivatives of phosphinic acid (*17*→*19*); however, the possibility of competing insertion of carbenes into the H/X bond of the additives is always present, giving phosphine oxides with X in the α-position (*16*→*18*). Reaction branching at the carbene *16* was first observed on photolysis of *7* in water [13] and prompted detailed investigations on the phosphorylcarbene/methyleneoxophosphorane rearrangement.

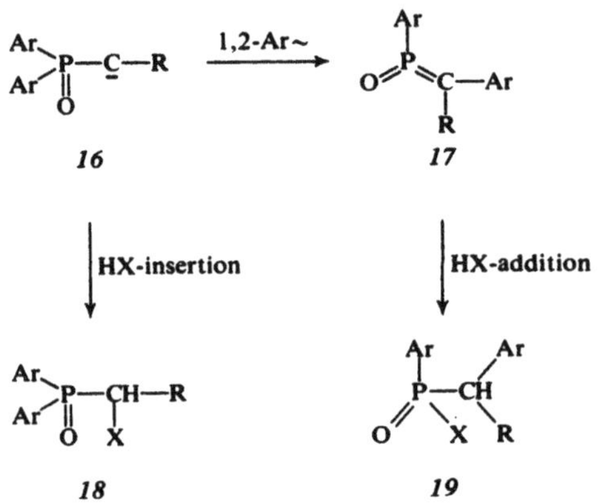

Irradiation of *7* in methanol gives a quantitative yield of α-methoxyphosphine oxide *18a* [11,14]; however, when the same carbene (*16a*) is generated by thermolysis of *7* in ethanol then rearrangement to *17a* also takes place [15]. In view of the behavior of *16a*, para substituents in the phenyl groups of the phosphoryl group influence the rearrangement only if they exert a promoting effect. Thus it is understandable that

Table 1. Insertion reaction and phosphene rearrangement of diarylphosphorylcarbenes (*8*)

18, 19	Ar	R	X	% *18*	% *19*
a	C_6H_5	C_6H_5	OCH_3	100	—
b	C_6H_4—Cl—(p)	C_6H_5	OCH_3	78	—
c	C_6H_4—OCH_3—(p)	C_6H_5	OCH_3	52	26
d	C_6H_5	H	OH	—	84
e	C_6H_5	H	OCH_3	—	61
f	C_6H_5	H	morpholino	—	54

chlorine-substitution does not lead to phosphinic ester formation (*19b*), whereas methoxy-substitution leads to reaction branching at the carbene *16c* which gives *18c* and *19c* in a ratio of 2:1 (Table 1) [16]. Thus the 1,2-shift of the Ar groups of *16* to the electron-deficient carbene carbon is promoted by donor substituents.

In complete contrast, the photolysis of (diazomethyl)diphenylphosphine oxide completely avoids the insertion (*16*→*18*). High yields of the corresponding phosphinic acid derivatives (*19d–f*) are found both in water and methanol and in the presence of morpholine (see Table 1) [11,14]. In general, methyleneoxophosphoranes show the same reactivity towards protic nucleophiles as other heterocumulenes.

Remaining doubts as to the existence of methyleneoxophosphoranes of type *9* could be dispelled by trapping reactions with suitable cycloaddends, preference generally being given to carbonyl compounds. Of course, the unavoidably generated carbene *8* inevitably takes its toll. Thus the carbene *8* generated on irradiation of *7* in benzene in the presence of aldehydes undergoes up to 23% [2 + 1]-cycloaddition with the solvent, giving the norcaradiene *20* whose stereochemistry at C-7 has been unequivocally determined by ^{1}H-NMR spectroscopy; insertion into the aldehydic C/H-bond is also observed [17,18].

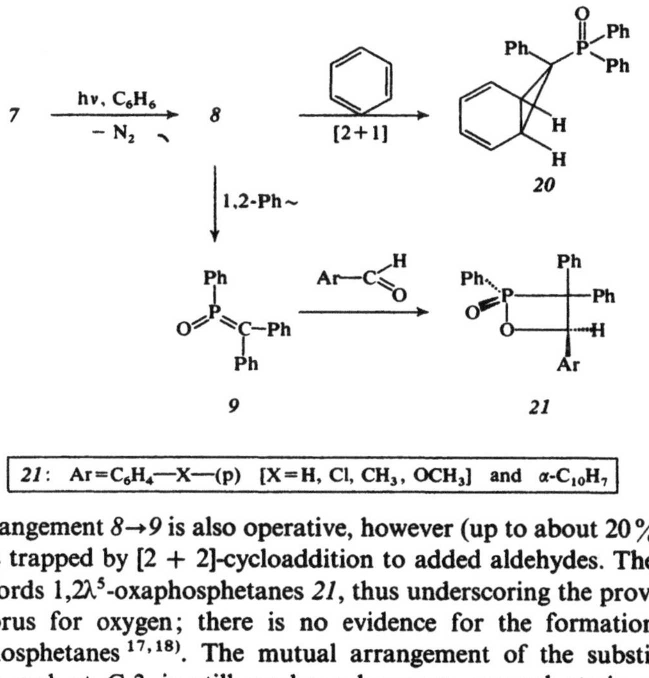

21: Ar=C_6H_4—X—(p) [X=H, Cl, CH_3, OCH_3] and α-$C_{10}H_7$

The rearrangement *8*→*9* is also operative, however (up to about 20%); the heterocumulene is trapped by [2 + 2]-cycloaddition to added aldehydes. The regiospecific reaction affords 1,2λ^5-oxaphosphetanes *21*, thus underscoring the proverbial affinity of phosphorus for oxygen; there is no evidence for the formation of isomeric 1,3λ^5-oxaphosphetanes [17,18]. The mutual arrangement of the substituents at the phosphorus and at C-3 is still unclear; however, general steric considerations suggest that the aromatic groups in these positions will be trans oriented [18]. Hydrolysis of *21* (giving β-hydroxyphosphinic acids) distinguishes the 1,2-oxaphosphetanes from the 1,3-isomers just as clearly as does their fragmentation in the mass spectrometer, yielding the characteristic fragments $M^+ - PhPO_2$ in relatively high intensity (39–100%) [18].

α,β-Unsaturated aldehydes and ketones likewise initially scavenge the methyleneoxophosphorane *9*, generated photolytically from *7*, to form 1,2λ^5-oxaphosphetanes

(*22*); i.e. the carbon-carbon double bond is not immediately involved in the reaction [18,19,20].

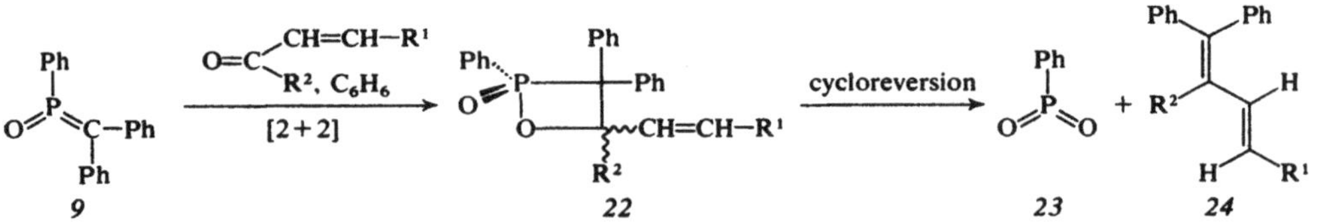

22, 24	*a*	*b*	*c*	*d*	*e*
R^1	C_6H_5	C_6H_4—OCH_3—(p)	C_6H_4—CH_3—(p)	C_6H_5	CH=CH—C_6H_5
R^2	H	C_6H_5	C_6H_5	C_6H_5	C_6H_4—CH_3—(p)

In the case of *22b–e*, the butadienes *24b–d* and the hexatriene *24e*, respectively, are also obtained; on trapping with cinnamaldehyde, the 1,3-diene *24a* is even the sole reaction product. It is quite obvious that the olefins *24* are secondary products of the trapping reaction of *9* arising by photofragmentation of *22*. The other product is phenyldioxophosphorane (*23*) which also numbers among the short-lived compounds of quinquevalent phosphorus with coordination number *3* (see Sect. 3.1).

The fact that only the vinyl-substituted 1,2λ^5-oxaphosphetanes *22* but not the arylated phosphorus heterocycles *21* undergo photofragmentations is presumably due to the inability of the latter to absorb the light (λ > 280 nm) supplied for carbene formation (*7*→*8*) [e.g. *21*, Ar = C_6H_5: $\varepsilon_{280} \approx 200$; *22b*, *d*: $\varepsilon_{280} \approx 9000$ (in methanol)] [18,20].

Typical carbene reactions, such as norcaradiene formation with solvent (*20*) must be accepted, as must the occasional cyclopropanation of the trapping reagents.

Viewed systematically, formation of the olefins *24* on reaction of methyleneoxophosphorane *9* with α,β-unsaturated carbonyl compounds is to be classified as an olefination reaction. The similarity to the Wittig reaction is obvious, the differences being just a matter of degree.

The generally quite stable methylenephosphorane nevertheless resembles the short-lived highly reactive methyleneoxophosphorane. The oxaphosphetane intermediate *25* formed by [$\pi^2_s + \pi^2_a$]-cycloaddition, which can only be isolated in exceptional cases [21], is to be seen against the stable oxaphosphetanes of type *26*, which can be photolyzed if suitably substituted or thermolyzed under drastic

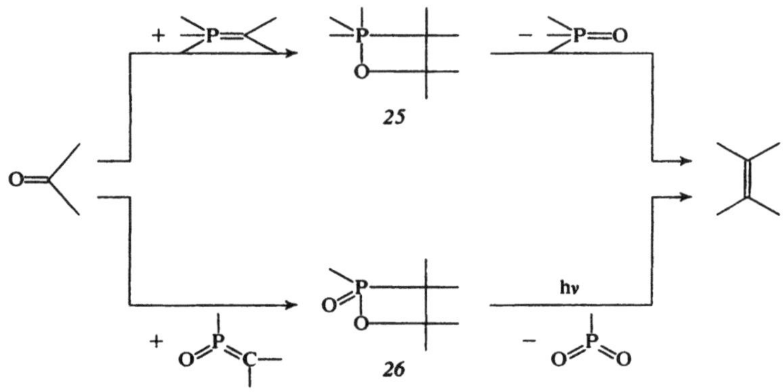

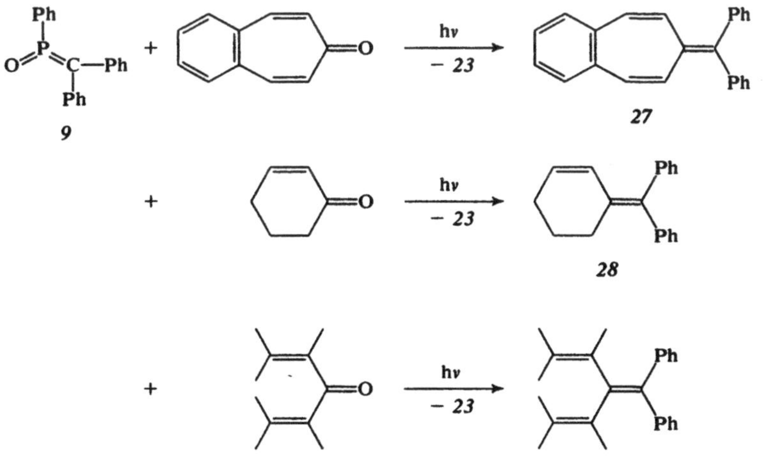

9 27 28 29

conditions. Thus on vacuum pyrolysis (150–160 °C/15 torr) the oxaphosphetane *21* with Ar = C_6H_5 undergoes cycloreversion to *23* and triphenylethylene, which subsequently isomerizes to 9-phenyl-9,10-dihydroanthracene under thermal conditions [22]. Finally, the fragments formed on decomposition of the heterocyclic intermediates *25* and *26* also resemble each other: the counterpart of the stable triphenylphosphine oxide is the highly reactive phenyldioxophosphorane *23*.

Suitable carbonyl compounds can thus be olefinated photochemically with (diazobenzyl)diphenylphosphine oxide (*7*), the oxygen function being substituted by a diphenylmethylene group [18,20]. Hence irradiation of *7* for a sufficient length of time in the presence of the corresponding unsaturated ketones affords the heptafulvene *27* [23], the trans-1,3-butadiene *28* [22], and the cross-conjugated hexatriene *29* [22] by direct olefination with the intermediate *9*.

A somewhat different result is obtained on reaction of methyleneoxophosphorane *9* (generated thermally from *7*) with α,β-unsaturated ketones. Owing to the unusual thermal stability of *7* [13,24] the reaction is carried out in a melt of the trapping reagents at 125 °C. Not surprisingly, the same product spectrum as for photolysis is initially observed, i.e. formation of oxaphosphetanes *22b–d* (14–26%) and of 1,3-butadienes *24b–d* (12–27%). The products, however, are accompanied by 3,4-dihydro-1,2λ^5-oxaphosphorins *30a–c* (14–30%) arising by hetero-Diels-Alder reaction between *9* and trapping reagent [19,20].

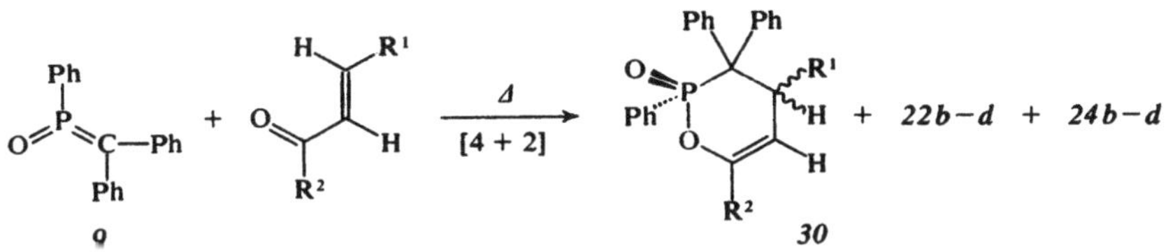

9 30

30	a	b	c
R¹	C_6H_4—OCH_3—(p)	C_6H_4—CH_3—(p)	C_6H_5
R²	C_6H_5	C_6H_5	C_6H_5

The product profile thus reveals impressive parallels with the reaction of diphenylketene, the carbon analogue of *9*, with (p-methoxybenzal)acetophenone, in which, again under thermal conditions, both cycloadditions and fragmentation of the four-membered ring product [25] occur. Overall, the rate or rearrangement *7*→*9* appears to be more favorable by the thermal route than by the photochemical pathway.

Several unusual cycloaddition reactions of *9* with unsaturated ketones should be mentioned in conclusion: the heterocumulene generated photolytically from *7* undergoes [8 + 2]-cycloaddition with tropone to form *33* (40%); the structure of the product has been unequivocally established by X-ray structure analysis [22,23]. Once again, the affinity of phosphorus for oxygen is manifested; an entirely analogous cycloaddition reaction is known for diphenylketene [26].

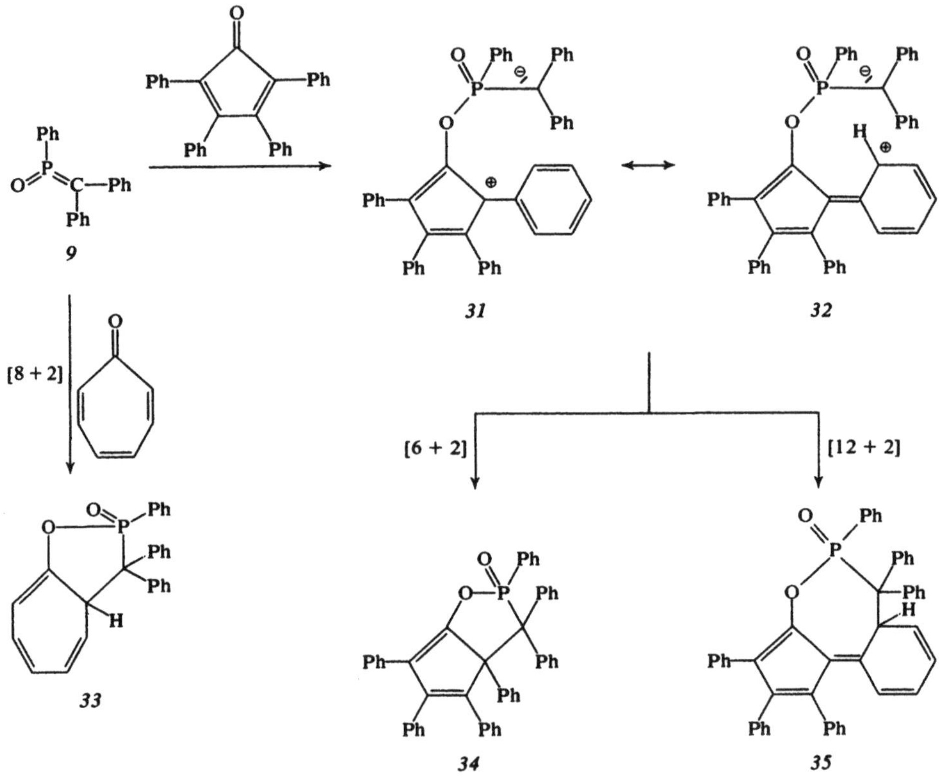

The reaction of *9* (generated thermally from *7* in toluene) with tetraphenylcyclopentadienone is more complex. Both the [6 + 2]-cycloadduct *34* [16], for which an X-ray structure analysis is available, and the [12 + 2]-cycloadduct *35* [16], whose constitution has been assigned primarily on the basis of ^{1}H-NMR evidence, are obtained. The two cycloadducts presumably have a common intermediate which, in accord with the general reactivity of *9*, should possess betaine character (*31*↔*32*); it is caused by nucleophilic attack by the carbonyl oxygen atom on the phosphorus of the heterocumulene. Ring closure of the carbanionic carbon atom with the electrophilic centers shown in formulae *31* and *32* then leads to the two isomeric P/O heterocycles.

The other [2 + 2], [4 + 2], and [8 + 2]-cycloaddition reactions of *9* might proceed via analogous two-step mechanisms.

2.2 Acylmethyleneoxophosphoranes

Like the arylmethyleneoxophosphoranes *9*, the acylmethyleneoxophosphoranes *39* have so far resisted isolation as such. Flash pyrolysis [12] of (diazophenacyl)-diphenylphosphine oxide (*36a*) [27] is just as unsuccessful as that of *7*; the sole product isolated is diphenylacetylene which is presumably formed via *39* or the cyclic isomer 1,2λ^5-oxaphosphetane by subsequent fragmentation of *23*.

A more successful approach is the photolysis of *36*, in which the acylmethyleneoxophosphoranes *39* formed from *37* by 1,2-phenyl shift can be trapped [10,11,14]. It should, however, be borne in mind that the carbene intermediate *37* not only undergoes the phosphene rearrangement, but "classical" Wolff rearrangement can also occur to give the phosphorylketene *41*. If the photolysis is performed in methanol, then both heterocumulenes are consequently transformed into the acid derivatives *42* and *44*. That insertion of the carbene *37* into the O/H bond of methanol, giving *40*, competes with the two rearrangement reactions is just as unavoidable as photochemical isomerization of product *40* to the oxete *43* [11]; the extent of this reaction depends upon the duration of irradiation. Reductive elimination of nitrogen [11,28,29] from *36* to form *38*, whose mechanism is not under discussion in this context, finally rounds of the product scenario. Acyl substituents R might plausibly affect the course of rearrangement (see Table 2), as might donor and acceptor substituents in the para position of the PO-phenyl group.

In the case of R = C_6H_5, the P/C phenyl shift dominates over the C/C shift to the extent of 2:1, the statistical factor already having been accounted for [11].

Table 2. Insertion reaction, phosphene and ketene rearrangements of phosphorylacylcarbenes (*37*)

36–44	R	% *42* (P/C—Ar~)	% *44* (C/C—Ar~)	% *40**	% *38*
a	C_6H_5	44	12	13	5
b	4-pyridyl	40	8	2	21
c	2-furyl	85	—	10	—
d	2-thienyl	65	11	—	—
e	2-benzofuryl	69	—	5	—
f	2-benzothienyl	52	—	36	—
g	2-pyrroyl	21	—	29	32
h	9-anthryl	—	85	—	—
i	1-naphthyl	33	37	14	—
j	mesityl	—	40	—	—
k	4-chinolyl	30	32	—	—
l	tert-butyl	4	—	40	25
m	1-adamantyl	9	—	56	—

* Yield of *43* included.

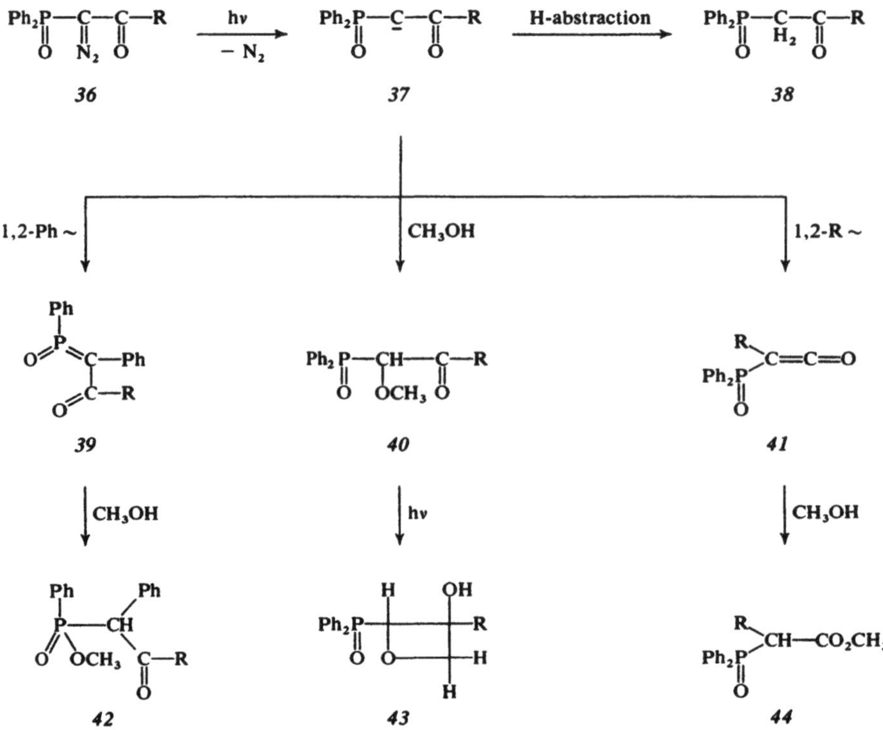

p-Methoxy substitution in the phosphoryl group again raises the yield of *42* (60%), while p-chloro substitution of course lowers it (27%) [16]; such substitution is without effect upon the extent of Wolff rearrangement (13%). Increase and decrease of phosphene rearrangement is reflected in the extent of O/H insertion, i.e. formation of products *40* and *43*. Groups R have a direct effect on the Wolff rearrangement. While the 4-pyridoylcarbene *37b* behaves similarly to *37a*, at least with regard to P/C-phenyl and C/C-pyridyl shifts (formation of *42b* and *44b*) (Table 2), the marked influence of the other substituents can be interpreted as follows [30]:

1) 2-Furyl, thienyl, benzofuryl, benzothienyl, and pyrroyl groups in *37c–g* prevent, or largely suppress Wolff rearrangement (see Table 2). This is due to the donor character of the heteroaromatic moiety, which endows the heteroaryl/CO bond with partial double bonding character and thus hinders rearrangement; the amount of phosphinic esters *42c–g* is correspondingly high [30].

2) 9-Anthryl, 1-naphthyl, mesityl, and 4-quinolyl groups in *37h–k* promote C/C aryl shift (see Table 2); their proclivity to migrate is far greater than expected from the electronic influence of the π-systems. This is presumably due to the intermediacy of spiro-linked benzenium betaines [30, 31], occurring in place of the carbenes otherwise encountered. The formation of phosphinic esters *42h–k* is partly or completely suppressed [30].

3) Tert-butyl and 1-adamantyl groups in *37l* and *m*, respectively, completely suppress Wolff rearrangement for general steric reasons (see Table 2), without the phosphene rearrangement to *42l* and *m* playing any significant role; preferential O/H insertion occurs with the solvent [30]. The likelihood for phosphorylcarbene/

methyleneoxophosphorane rearrangement should increase, for statistical reasons if no other, if the diazoalkane is substituted on both sides by diphenylphosphoryl groups. Thus, the photolysis of diazo bis(diphenylphosphoryl)methane (*45*) in dioxane/water proceeds via the carbene and phosphene intermediate (*46* and *47*, respectively) to give the phosphinic acid *48a* in a yield of 92%, previously unknown for this rearrangement[32].

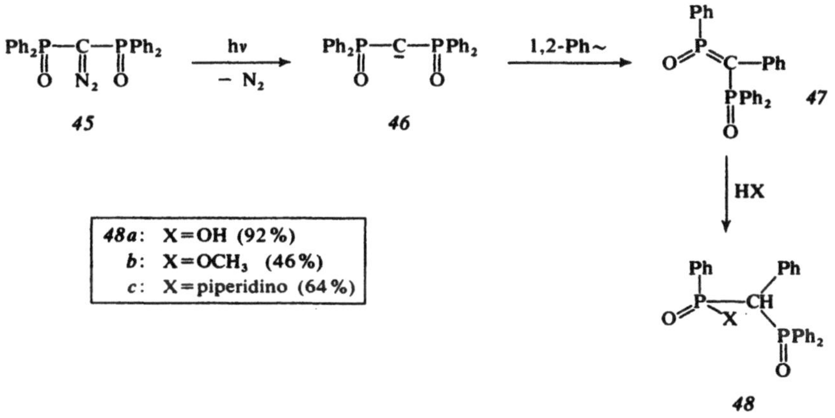

The same reaction sequence performed in methanol affords a mixture of diastereomers of the phosphorylated phosphinic ester *48b*, of which one pure isomer can be isolated[32]. In the presence of piperidine, reductive elimination of nitrogen[28,29] from *45* to give bis(diphenylphosphoryl)methane competes with the prevailing formation of the phosphinic piperidide *48c*[32]. Expected trapping of *47* by [2 + 2]-cycloaddition with benzaldehyde fails to occur: in place of 1,2λ⁵-oxaphosphetanes, products are obtained which arise mainly by way of the benzoyl radical[32,33].

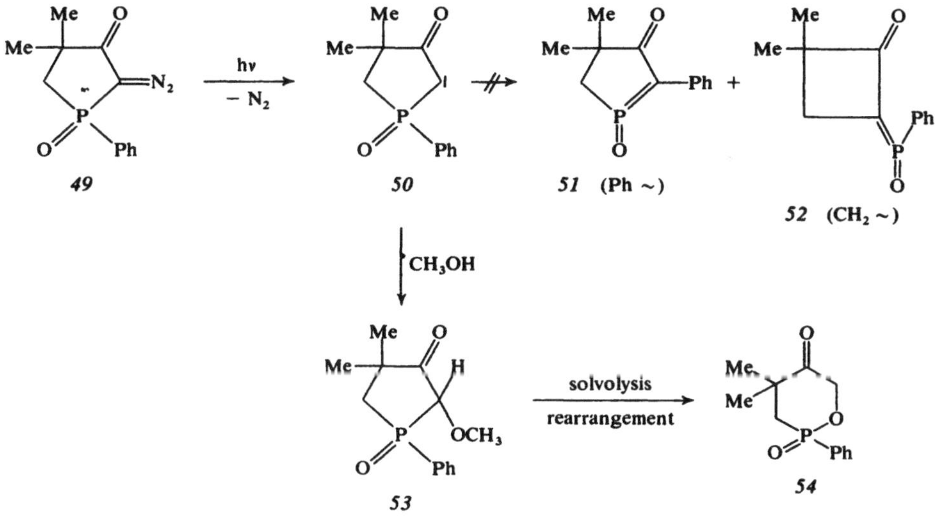

Photolysis of the first known "cyclic" α-diazo-β-oxophosphine oxide *49* is unsuccessful with regard to phosphene formation. There is no evidence for a P/C-phenyl shift, which should lead to *51*, nor for a P/C-alkyl shift, which would afford *52* via ring contraction, since none of the expected phosphinic esters could be isolated in methanol [34].

Both reactions are prevented on the one hand because the diazo compound undergoes solvolysis with methanol, with cleavage of the PO/CN_2 bond; the newly formed α-diazo ketone consequently undergoes Wolff rearrangement [34]. On the other hand, rearrangements are circumvented by reductive nitrogen elimination [28,29] and by methanol insertion to give *53*. Although the latter product could not be isolated, the 1,2λ⁵-oxaphosphorinane *54* obtained would appear to have arisen from *53* by hydrolysis and rearrangement [34].

Trapping reactions of benzoylmethyleneoxophosphorane *39a* with carbonyl compounds dispel any remaining doubts as to the existence of acylated phosphenes. Unlike the diphenylmethyleneoxophosphorane *9*, whose P/C double bond participates in cycloadditions, compound *39a* acts as a hetero-1,3-diene and undergoes [4 + 2]-cycloaddition with aldehydes and ketones [10,17,35]; it may again be assumed that the reaction is a two-step process involving *55* as intermediate.

Irradiation of *36a* in 1,4-dichlorobenzone at 60 °C, a temperature at which thermal decomposition of the diazo compound is still negligible [24], in the presence of benzophenone, acetone, or cyclohexanone leads to 1,3,4λ^5-dioxaphosphorins *56a–c* (≡ *57a–c*) [35].

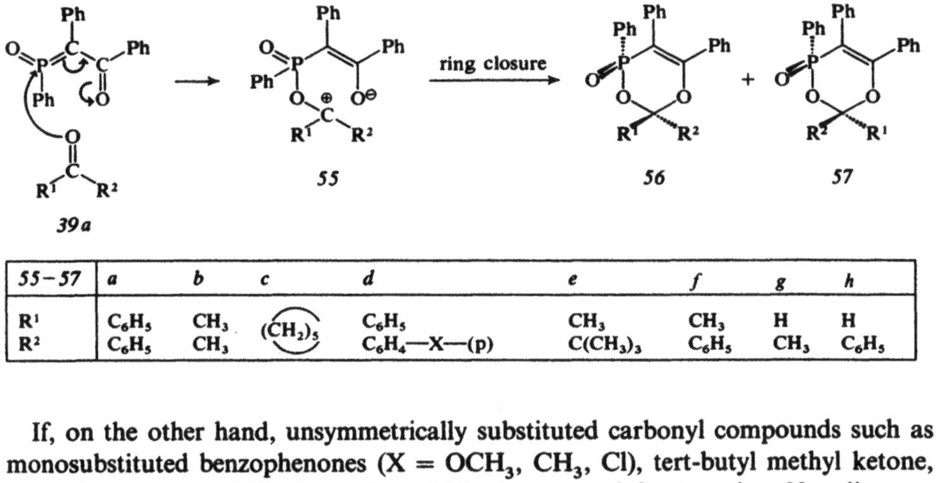

55−57	a	b	c	d	e	f	g	h
R¹	C_6H_5	CH_3	$(CH_2)_5$	C_6H_5	CH_3	CH_3	H	H
R²	C_6H_5	CH_3		C_6H_4—X—(p)	$C(CH_3)_3$	C_6H_5	CH_3	C_6H_5

If, on the other hand, unsymmetrically substituted carbonyl compounds such as monosubstituted benzophenones (X = OCH_3, CH_3, Cl), tert-butyl methyl ketone, acetophenone, acetaldehyde, or benzaldehyde are used for trapping *39a*, diastereomeric mixtures are formed in each case; they could all be resolved except for the products obtained with p-methoxybenzophenone and acetophenone [35]. An X-ray structure analysis has been performed for the E-isomer *57g* [36] which, in conjunction with ¹H-NMR studies, permitted structural assignment in cases *56* and *57e*, *g* and *h* [35]. Additional chemical evidence for the structure of the six-membered heterocycles is provided by the thermolysis of *56a* considered in another context (see Sect. 3.1). In general the reaction *39a*→*56* or *57* is accompanied by formation of phosphene dimers, presumably via [4 + 4]- and via [4 + 2]-cycloaddition [35].

Compound *39a* reacts with N-methylbenzaldehyde imine as cycloaddend under the conditions metioned above — quite expectedly — to give a mixture of stereoisomeric 1-oxa-3-aza-4λ^5-phosphorins [37].

2.3 Aryl- and Acylmethylenethiophosphoranes

Comparatively little is known about the sulfur analogues of *9* and *39a*, i.e. the methylenethiophosphoranes *59a* and *b*; in both cases they have been generated from appropriate α-diazothiophosphine oxides by methods based on previous results.

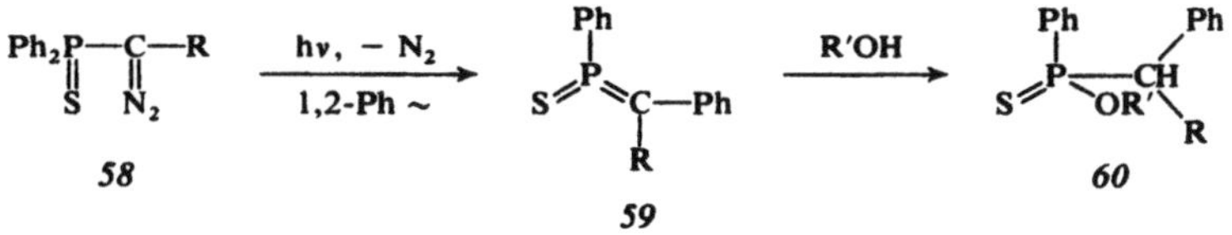

58–60 *a*: R=C_6H_5; *b*: R=COC_6H_5

Thermolysis of *58a* in butanol affords, together with 17% of *60a* (R′ = C_4H_9) which evidences the intermediacy of the thiophosphene *59a*, a variety of partly atypical products which seriously impede the desired rearrangement [38]. Photolysis of *58b* in methanol is also found to give only 18% 1,2-P/C shift to form the heterocumulene *59b*, from which the thiophosphinic ester *60b* (R′ = CH_3) results [39]. As already mentioned in connection with the photolysis of diazo compounds of type *36* (see Sect. 2.2), Wolff rearrangement (9%) and O/H insertion (6%) once again compete with thiophosphinic ester formation. Moreover, solvolysis of the P(S)/C(N_2) bond [39] prevents a greater contribution of carbene products to the overall yield.

3 Metaphosphonates

Aryldioxophosphoranes such as aryl- and alkyl-iminooxophosphoranes number among the short-lived metaphosphonates. The former are best obtained by fragmentation of cyclic phosphinic esters, and the latter by rearrangement of aryl- and alkyl-substituted phosphoryl azides and nitrenes, respectively. This reaction is reminiscent of the phosphorylcarbene/methyleneoxophosphorane rearrangement discussed in Section 2.

3.1 Aryldioxophosphoranes

The preparation of phenyldioxophosphorane ("benzenephosphonic anhydride") was reported as long ago as the turn of the century. It was thought to arise on reaction of benzenephosphonic acid with benzenephosphonic dichloride [40]; aryl-substituted derivatives were also believed to have been formed by the same reaction [41].

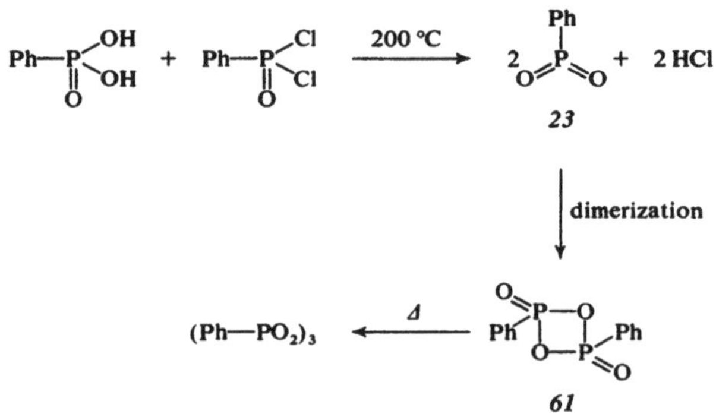

Meanwhile, however, this reaction has been found to give mainly the dimeric metaphosphonate *61*; since this product yields a trimer on heating to above its melting point (103–105 °C) we can justifiably assume that the above reaction, performed at 200 °C, likewise affords the trimer of *23* rather than *23* itself[42].

A potential source for generating monomeric *23* is found in the 1,2λ^5-oxaphosphetanes *21* and *22*[18,20]. Their mass spectra contain peaks at M$^+$ — 140, corresponding to [2 + 2]-cycloreversion to olefin and *23*; however, the latter fragment (m/e = 140) was not found in the mass spectra. Although it cannot be explicitly stated whether this fragmentation is induced by electron-impact or thermally, a thermal reaction in the mass spectrometer certainly appears plausible. Such a reaction can indeed be accomplished on a preparative scale under milder conditions, as previously reported for *21* (R = C_6H_5) (Sect. 2.1).

The photochemical fragmentation of vinyl-substituted 1,2λ^5-oxaphosphetanes, representing a step of a photochemical variant of the Wittig reaction with methyleneoxophosphoranes, has been examined as a model in the case of *22b*[20]. Photolysis of this compound in methanol affords the 1,3-diene *24b* as well as the highly reactive dioxophosphorane *23* which is trapped by the solvent; subsequent esterification of the half-ester *62*, formed as a primary product, with diazomethane to give the diester *63* was undertaken solely for preparative reasons[20].

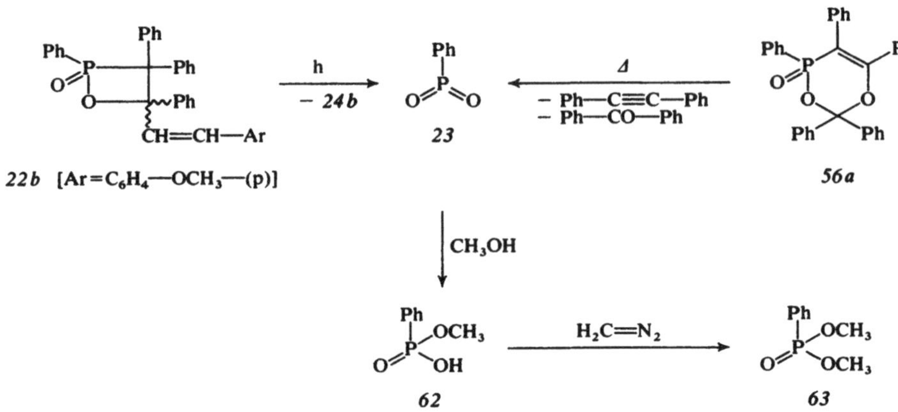

Formation of the metaphosphonate *23* also occurs on mass-spectrometric fragmentation of the 1,3,4λ^5-dioxaphosphorin *56a* [17,35]; a peak at m/e = 140 corresponding to the mass number of *23* is indeed observed [35]. Vacuum pyrolysis of *56a* parallels the mass spectrometric degradation to form diphenylacetylene and benzophenone; however, *23* could not be detected directly. This can be accomplished though, if thermolysis is conducted in methanol (220 °C/22 at); dimethyl benzenephosphonate (*63*) is again formed via *62* [35].

In view of the reaction behavior of 1,2λ^5-oxaphosphetanes (*22*), treated above, it appears fitting to reconsider the mechanism of the hydroxylion induced fragmentation of β-bromophosphinic acid *64* [43]. It was assumed that formation of the phosphinate *65* is followed by that of the four-membered heterocycle *66*, which spontaneously decomposes to benzalacetophenone and phenyldioxophosphorane; the latter then adds water to give the phosphonic acid [43].

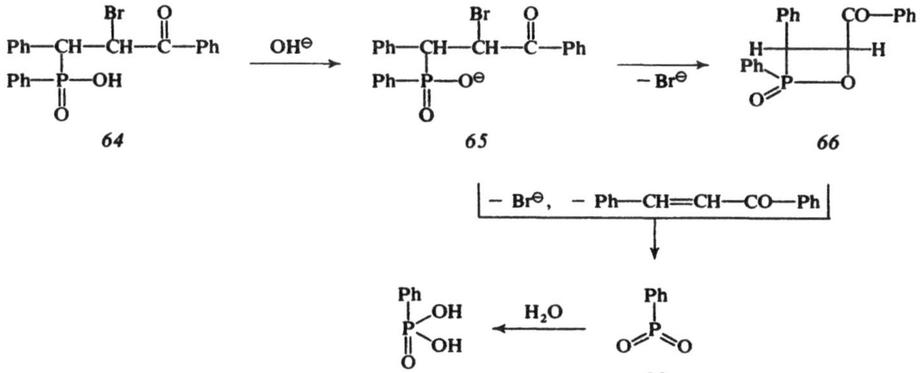

64 *65* *66* *23*

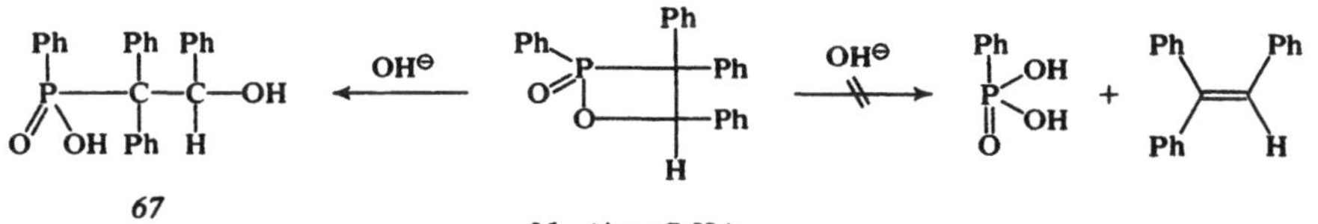

67 *21* (Ar=C_6H_5)

This concept is undoubtedly opposed by the fact that the structurally comparable 1,2λ^5-oxaphosphetane *21* (Ar = C_6H_5) undergoes only ring opening to the phosphinic acid *67* on alkaline hydrolysis and there is no evidence for the formation of benzenephosphonic acid and triphenylethylene [18].

Further work on generation of the highly reactive dioxophosphoranes utilizes the O/P-heterocycles *68* and *70* and is aimed at mesityldioxophosphorane *69* as target species [44] (see also Sect. 4.2 with regard to the method). High vacuum pyrolysis of *68* at 600 °C generally leads via a retro-Diels-Alder reaction to 1,3-butadiene and presumably to the metaphosphonate *69*. Contrary to expectation, the ^{31}P-NMR spectrum of *69* condensed directly from the pyrolysis zone into a trap cooled to ca. −80 °C shows a complex splitting pattern which simplifies to a singlet at δ = 12.4 ppm only on warming to room temperature. Treatment of this, presumably polymeric, metaphosphonate with aqueous sodium hydroxide then furnishes mesitylphosphonic acid [44].

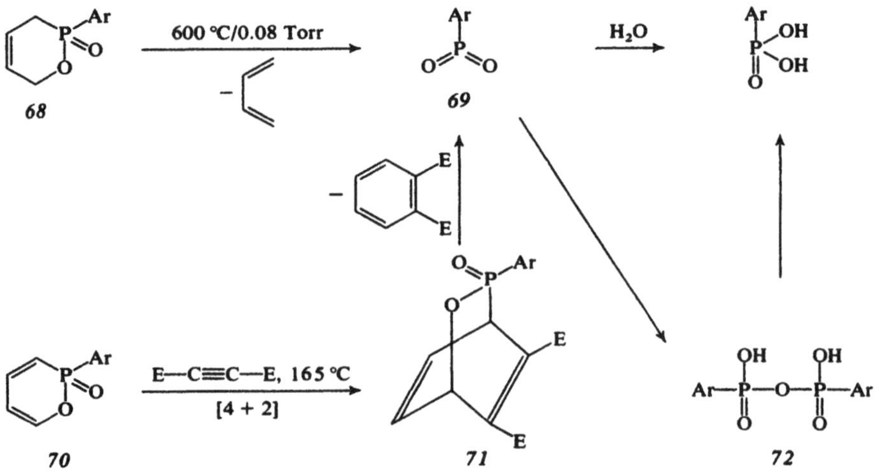

Ar = mesityl; E = $CO_2C_2H_5$

Generation of highly reactive mesityldioxophosphorane (*69*) is also accomplished on subjecting the 1,3-diene *70* to a Diels-Alder reaction with diethyl acetylenedicarboxylate at 165 °C. The cycloadduct *71* cannot be detected directly; on work-up of the reaction mixture with aqueous sodium hydrogen carbonate it is, however, possible to isolate the pyrophosphonate *72* which is cleaved to give mesitylphosphonic acid in more strongly alkaline medium [44].

3.2 Aryl- and Alkyliminooxophosphoranes

As already mentioned, rearrangement of phosphoryl azides or corresponding nitrenes represents a method of generating iminooxophosphoranes; it has been studied in detail for 1-azidophosphetane 1-oxides [45,46,47]. The results obtained with 1-azido-2,2,3,4,4-pentamethylphosphetane 1-oxide (*78*) are presented as a model case; the photolysis of this compound in methanol has been examined very thoroughly with the aid of high pressure liquid chromatography [47]. Starting from trans-phosphoryl azide *78* nine products were isolated and identified, the reactive intermediates *79*, *80*, and *83* being seen to play a key role. The rearrangement process of primary interest starts from *80* and concludes with the formation of cis/trans isomeric cyclic phosphonic amidic esters *81* and *82* (30:70) and also represents the principal reaction [47]. Interestingly, the same product ratio results starting from the cis-phosphoryl azide. The original stereochemical distinction between the two reactants must have been lost at some stage of the reaction; this can be accounted for by a trigonal-planar heterocumulene intermediate (*80*) (a pyramidal structure with fast inversion cannot be excluded a priori). Nucleophilic azide exchange with methanol to give *73* plays only a minor role; this also applies in principle to C/H insertion

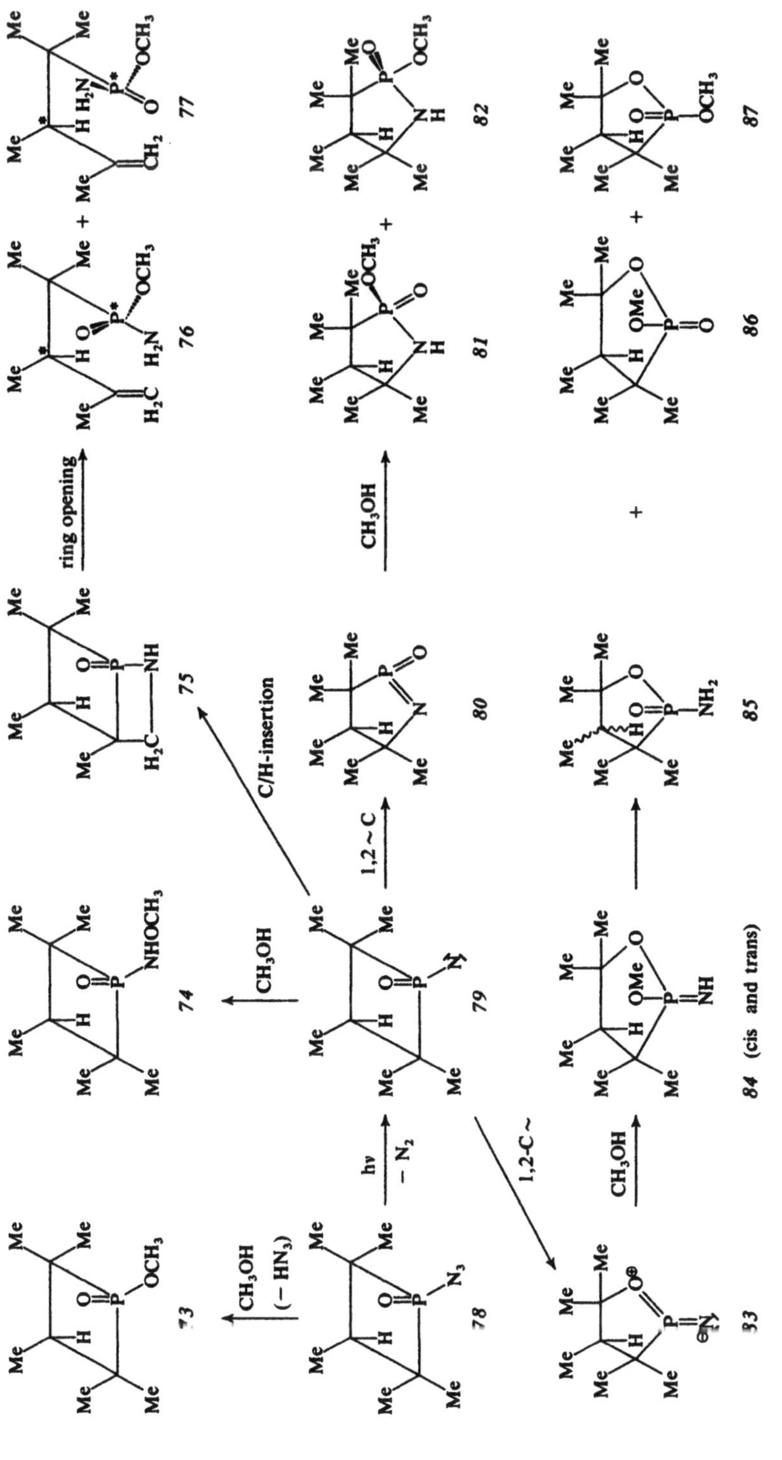

Yields:

81+82	73	76+77	74	85	86+87
65%	6%	14%	1%	2%	11%

of the phosphorylnitrene *79* which yields the amide *74*. Considerably more important is the formation of the open-chain phosphonic amidic esters *76* and *77* involving ring cleavage and rearrangement. These esters represent diastereomers and are formed stereoselectively (9:1); the ratio of diastereomers is inverted in the case of the cis-phosphoryl azide. The bicyclic intermediate *75* is more or less speculative [47].

A significant contribution to the product pattern is made by the 1,2-C shift at the nitrene stage *79*; the shift does not take place to the electron-deficient nitrene nitrogen but to the oxygen atom of the phosphoryl group, and affords an intermediate which derives, like *80* from quinquevalent-tricoordinate phosphorous; it is designated as "metaphosphazile". The usual addition of methanol affords the cis/trans phosphonoimidic esters *84* which give compounds *85*–*87* on further reaction with methanol or on chromatographic work-up [47].

Similar results are obtained with 1-azido-2,2,3,3-tetramethyl- and 1-azido-2,2,4,4-tetramethylphosphetane 1-oxide on photolysis in methanol [45,46]. The fact that the unsymmetrically substituted phosphoryl azide undergoes both conceivable rearrangements, i.e. 1,2-shift of the CH_2 and the $C(CH_3)_2$ group, warrants mention.

Attempts to initiate formation of a nitrene, and its rearrangement to the iminooxophosphorane *80*, by subjecting 1-chloroamino-2,2,3,4,4-pentamethylphosphetane 1-oxide to α-elimination with sodium methoxide proved unsuccessful [48]. In contrast, however, the phosphorylhydroxylamides *88* rearrange in the presence of tert-butylamine to the heterocumulene *89* and then add base to give the phosphonic diamides *90* (>90%) [49]. The reaction is reminiscent of the well-known Lossen degradation.

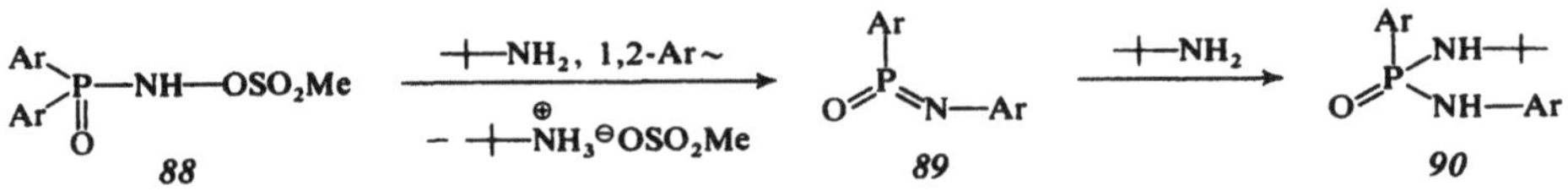

88 *89* *90*

Ar = C_6H_5, C_6H_4—CH_3—(p), C_6H_4—OCH_3—(p)

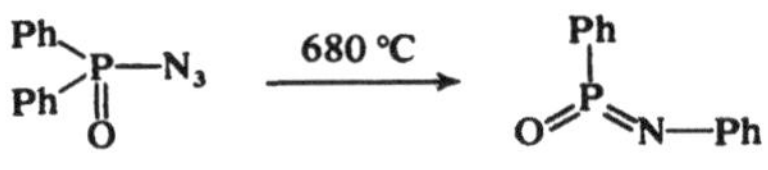

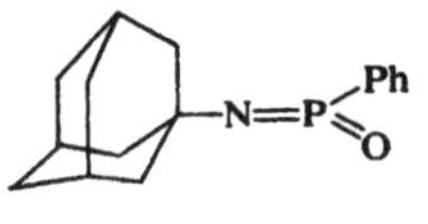

91 *92* *93*

Two other results will now be pointed out which presumably also require reinterpretation in the light of the reaction behavior of iminooxophosphoranes. Thus the gas phase pyrolysis of diphenylphosphoryl azide is reported to give monomeric *92* [50] and the dehydrohalogenation of phenylphosphoric adamantylamidic chloride with methylhydrazine the heterocumulene *93* [51], which is even considered resistant to water. Since partly correct analytical values are available, *92* and *93* may well be oligomers.

4 Metaphosphates

The monomeric metaphosphate ion itself commands a fair amount of attention in discussions of metaphosphates. It is postulated as an intermediate of numerous hydrolysis reactions of phosphoric esters [52,54,55] and also of phosphorylation reactions [56]; kinetic and mechanistic studies demonstrate the plausibility of such an assumption. In addition, the transient formation of ester derivatives of metaphosphoric acid — in which the double-bonded oxygen can also be replaced by thio and imino — has also been observed; they were detected mainly on the basis of the electrophilic nature of the phosphorus.

4.1 The Metaphosphate Ion

The monomeric phosphate ion *102* was first postulated in 1955 as an intermediate of the hydrolysis of monoesters of phosphoric acid in an aqueous medium [57,58]. Another 24 years were to elapse before compound *102* was observed directly, and then not in solution but in the mass spectra of some pesticides. The negative ion CI spectra of enol phosphates *94* and of the thiophosphoric ester *95* display an intense peak at m/e = 78.9590, which is unequivocally assigned to the $PO_3^{\ominus}$ ion [59].

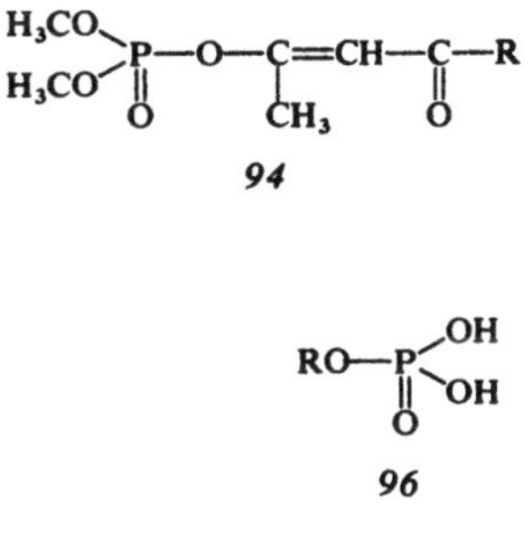

94

R = $NHCH_3$, OCH_3

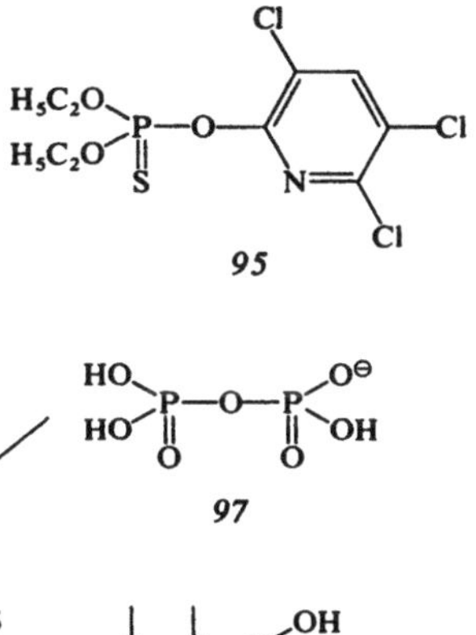

95

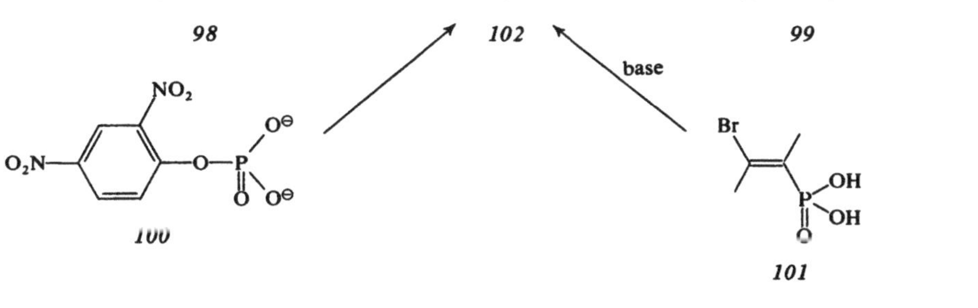

The fact that salts or other derivatives of monomeric metaphosphoric acid HPO_3 hitherto resisted isolation is generally attributed to the fast reaction of nucleophiles with the phosphorus of *102*. This also applies to its generation from derivatives of phosphoric and phosphonic acids *96–101*.

4.1.1 Hydrolysis of Monoesters of Phosphoric Acid

Many alkyl and aryl monoesters of phosphoric acid exhibit a characteristic dependence of their rate of hydrolysis upon the pH value of the solution as shown in Fig. 1 [58 b)].

The rate maximum at pH 4 is assigned to a specific reaction of the monoester anion *104* which exists exclusively under these conditions. Westheimer [57)] first advanced a metaphosphate ion mechanism in which *102* is formed via a six-membered monoester-anion/water complex (*103*). An intramolecular proton transfer via a four-membered ring according to *105* [60)] is also conceivable, as is the formation of a zwitterion *106* in a prior protonation equilibrium.

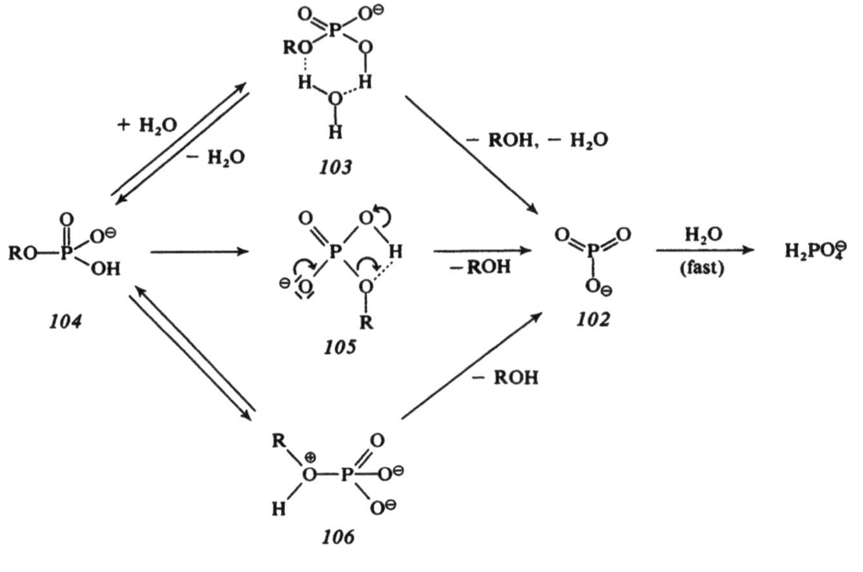

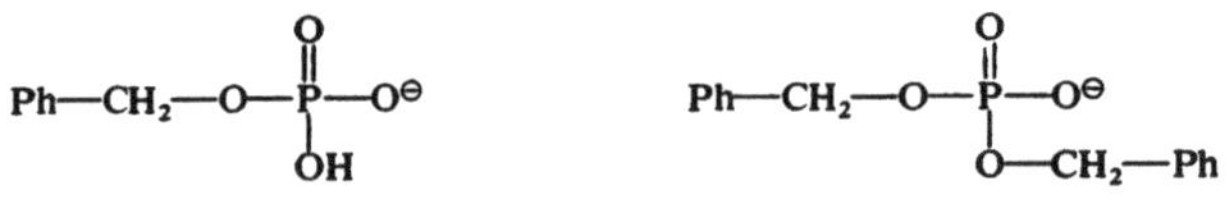

107 *108*

However the proton transfer may occur, it is of decisive importance that the leaving group should be a neutral molecule and the rate-determining step should be unimolecular. The following observations support the metaphosphate mechanism:

a) At the same pH value the dimethyl phosphate monoanion is hydrolyzed ca. 10^4 times slower than the monoethyl phosphate monoanion. The increase in hydrolysis rate with decreasing pH demonstrates that the neutral molecule is being hydrolyzed in the range between pH 0.72 and 4.17, with extensive cleavage of the C/O bond [61)]. This alternative reaction is understandable since the di-

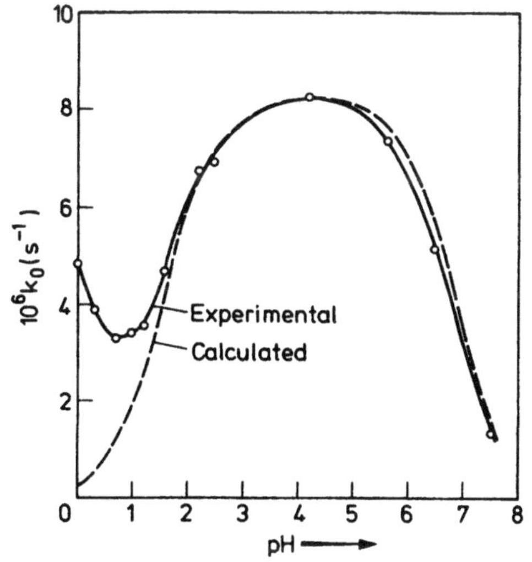

Fig. 1. pH-rate profile for hydrolysis of methyl phosphate at 100.1 °C [58b)]

methyl phosphate monoanion is unable to undergo intramolecular proton transfer.

b) Labeling experiments with $H_2^{18}O$ confirm the P/O bond cleavage in the course of hydrolysis of the monoanion [58 b)].

c) The rate constants and activation parameters of the hydrolysis of various alkyl, aryl, and pyranosyl monoesters according to *104* show little dependence upon the nature of the groups R [60)].

d) A linear free-energy relationship [62)] exists between log $k_{hydrol.}$ (1st order rate constant for the monoanion) and the pK_a value of the leaving group, whose slope (ca. —0.3) is compatible with departure of the neutral ROH but not with that of the high-energy $RO^{\ominus}$ ion.

e) The $\Delta S^{\neq}$ values lie close to 1 e.u. This is suggestive of a monomolecular rate determining step since, for example, the bimolecular hydrolysis of carboxylic esters is characterized by $\Delta S^{\neq}$ values of —20 e.u.

All arguments for the metaphosphate mechanism appear persuasive only when considered together since individual arguments could also apply to other hydrolysis mechanisms of phosphoric esters.

The pronounced proclivity of phosphoric monoester monoanions to eliminate $PO_3^{\ominus}$ is not always recognizable from the characteristic pH profile of Fig. 1. The hydrolysis rate maximum at pH ≈ 4 may be masked by a faster reaction of the neutral phosphoric ester, as in the case of α-D-glucose 1-phosphate [63)] or on hydrolysis of monobenzyl phosphate [64)]. In the latter case, the known ability of benzyl esters to undergo S_N1 and S_N2 reactions permits fast hydrolysis of the neutral ester with C/O bond breakage. The fact that the monoanion *107* of the monobenzyl ester is hydrolyzed some 40 times faster than the monoanion *108* of the dibenzyl ester at the same pH again evidences the special hydrolysis pathway of *107*, rationalized by means of the metaphosphate hypothesis.

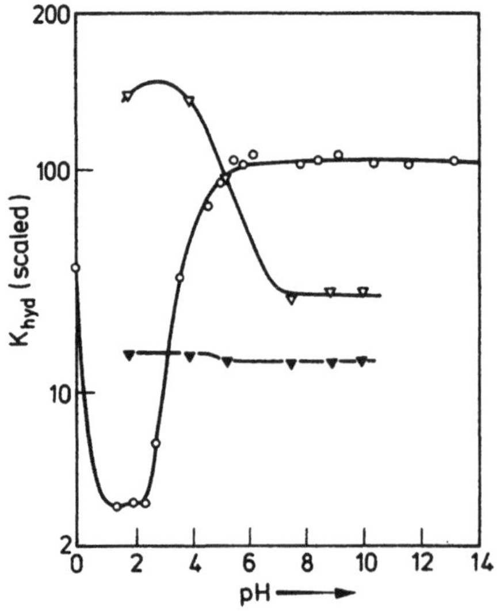

Fig. 2. Rate of hydrolysis vs. pH value (39 °C, ionic strength 1.0) for **a** 2,4,6-trichlorophenyl phosphate ($k_{hydrol.} \cdot 10^7$ min.$^{-1}$), **b** 2,4-dinitrophenyl phosphate ($k_{hydrol.} \cdot 10^4$ min.$^{-1}$), **c** 2-chloro-4-nitrophenyl phosphate ($k_{hydrol.} \cdot 10^6$ min.$^{-1}$). (Ref. [65])

As is apparent from Fig. 1, the dianions of monoalkyl phosphates normally resist hydrolysis. However, for leaving groups whose conjugate acids exhibit a $pK_a \lesssim 5$ in water, hydrolysis of the dianion becomes faster than that of the monoanion. Fig. 2 shows a pH profile characteristic of this situation. Whereas the hydrolysis rate of 2,4,6-trichlorophenyl phosphate (pK_a of the phenol 6.1) still shows the typical monoanion preference as seen for methyl phosphates (Fig. 1), the dianion of 2,4-dinitrophenyl phosphate (pK_a of the phenol 4.09) is hydrolyzed far faster than the monoanion; 2-chloro-4-nitrophenyl phosphate represents an intermediate case (pK_a of the phenol 5.45) [65].

The hydrolysis rate of the monoester dianions falls with decreasing pK_a value of the conjugate acid of the leaving group. A linear relationship was found between pK_a and $k_{hydrol.}$ for the hydrolysis of derivatives of phenols with pK_a 4.07–7.23 at 39 °C [65]. The slope of —1.23, slightly positive $\Delta S^{\neq}$ values, and insignificant isotope effects contraindicate bimolecular displacement by a water molecule, and strongly suggest unimolecular decomposition of the dianions *109* with formation of the metaphosphate ion *102*.

A kinetic isotope effect $^{16}O/^{18}O$ of 2% in the spontaneous hydrolysis of the 2,4-dinitrophenyl phosphate dianion, whose ester oxygen is labeled, suggests a P/O bond cleavage in the transition state of the reaction, and thus also constitutes compelling evidence for formation of the metaphosphate [66,67]. The hydrolysis behavior of some phosphoro-thioates (*110*) is entirely analogous [68].

For *110a* the rate of hydrolysis in water at 35 °C follows the pH profile shown in Fig. 1, i.e. with a pH maximum at 3–4 where the monoanion predominates, while the dianion reacts extremely slowly; in principle, *110b* behaves analogously but is more reactive than *110a*. In contrast, the dianion of *110a* is hydrolyzed 16.2 times faster than the monoanion; thus it is comparable with the 2,4-dinitrophenyl phosphate dianion which is hydrolyzed about 30 times faster than its monoanion at 39 °C [65,69]. A good leaving group is also present in *110c*, namely the 4-nitrothiophenolate group, which is able to depart from the molecule without prior protonation at the sulfur.

The previously mentioned linear free energy relationship between $k_{hydrol.}$ and the pK_a value of the laving group for the monoanions of monoalkyl and monophenyl esters of phosphoric acid and for esters with leaving groups of intermediate basicity is no longer obeyed by esters with leaving groups of $pK_a \lesssim 7.4$. The deviations become greater with falling pK_a value (see [65,69]).

Doubts have recently been expressed regarding the validity of the metaphosphate pathway for hydrolysis of the monoanion of 2,4-dinitrophenyl phosphate (*111*) [70,71,72] since the basicity of the 2,4-dinitrophenolate group is insufficient to produce a zwitterion corresponding to *106* or even a proton transfer via intermediates of type *103* or *105* (pK_a values in water: 4.07 for 2,4-dinitrophenol, ~1.0 and 4.6 for 2,4-dinitrophenyl phosphate). Instead, hydrolysis and phosphorylation reactions of the anion *111* are formulated via oxyphosphorane intermediates according to *114*.

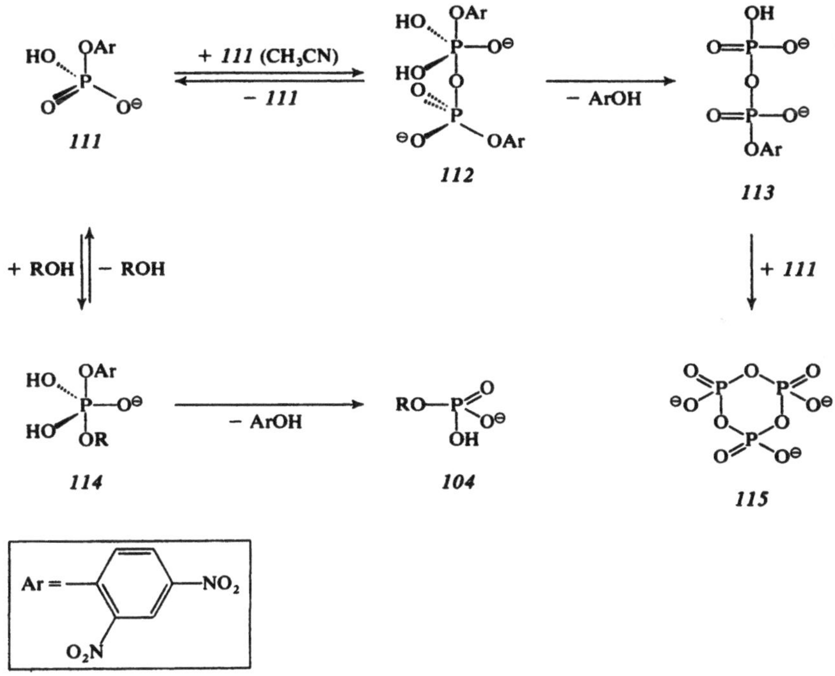

In the aprotic solvent acetonitrile, the trianion of cyclic trimetaphosphoric acid *115* is formed via the phosphorane *112* and the pyrophosphate dianion *113*; protic solvents differ by reacting via the phosphorane *114* to give the phosphates *104* or $H_2PO_4^{\ominus}$.

Evidence for the participation of oxyphosphorane intermediates in the reactions of monoanion *111* comes from:

a) "Cyclotrimerization" to *115* or phosphorylation of alcohols or water proceed at similar rates as for the free acid, considerably slower than the corresponding reactions of 2,4-dinitrophenyl phosphate dianion for which the metaphosphate mechanism is retained.

b) tert-Butanol is not phosphorylated by *111*, but it is phosphorylated by the analogous dianion; this is ascribed to the extraordinary reactivity of the monomeric metaphosphate anion.

The oxyphosphorane mechanism may be valid only for the case under study in which the counterions of the phosphate *111* are the tetra-n-butyl-ammonium and the diisopropylethylammonium cation, for which nucleophilic catalysis can be ruled out for decomposition of *111*. In contrast, a powerful influence on the hydrolysis rate of the same anion is found for a number of amine nucleophiles [73]. This, together with the rapidly decreasing influence of amine catalysis [73] with increasing pK_a value of the leaving group in phosphate monoester monoanions (powerful influence of leaving group and nucleophile) suggest a classical $S_N2(P)$ mechanism or participation of oxyphosphorane intermediates in which the amine component is incorporated [71,72].

Yet another situation is observed in the 2,4-dinitrophenyl phosphate dianion. A significant effect of amines on the rate of decomposition is admittedly observed; however, typical 2nd order kinetics, lower enthalpy of activation compared with spontaneous hydrolysis, and strongly negative ΔS^{*} values (see Table 3) indicate an $S_N2(P)$ reaction. Surprisingly, the reaction rate remains unaffected by the basicity of the amine, even when its pK_a value changes by 8 units.

It has to be assumed that these processes are occurring on the boundary between $S_N1(P)$ and $S_N2(P)$ mechanisms in whose transition states considerable P—O(—Ar) bond cleavage takes place. The lifetime of the resulting, more or less free metaphosphate anion *102* then depends upon the nucleophilicity of the surrounding solvent. With pyridine, for example, a very fast reaction occurs so that the overall process approaches an S_N2 reaction. Acceleration of the reaction by amines such as 2,6-lutidine, which are disqualified from acting as nucleophiles by steric hindrance, or by solvents such as dioxane, whiche are presumably too

Table 3. Activation parameters of spontaneous hydrolysis and second-order reactions with amines of the 2,4-dinitrophenyl phosphate dianion

Solvent	ΔH^{*} [kcal/mol]	ΔS^{*} [eu]	Ref.
Water	25.7	6.6	[65]
Pyridine, pH 7.1	15.5	−23.9	[66]
Pyridine, pH 9.2	16.8	−19.4	[73]
1,4-Dioxane, pH 9.2	23.1	− 3.8	[66]

weak to act as S_N2 nucleophiles [74], could be due to a solvent effect, "one possibility is solvation of the developing electrophilic centre, different only in degree of bond formation to phosphorus from the transition states for reaction with better nucleophiles" [73].

Generally speaking, it is possible that, owing to its extremely high reactivity, the metaphosphate ion (*102*) does not occur as a truly free species in any of its reactions, but is instead always "complexed" simultaneously by solvent or nucleophile and leaving group — a situation corresponding to a loosely coupled S_N2 transition state [75].

Acyl phosphates are also to be regarded as a potential source of metaphosphate (e.g. *98*→*102*). Both the acetyl phosphate dianion *116* and the acetyl phosphate monoanion *117* are amenable to elimination of low-energy leaving groups; that of the acetate ion in the first case, and that of a neutral acetic acid molecule after intramolecular proton transfer via a six-membered cyclic transition state in the second case. In contrast, the phosphoric diester monoanion *118* apparently does not undergo this decomposition mechanism, which would lead to the acetate ion and to phenyl metaphosphate comparable with *102*. The strongly negative $\Delta S^{\neq}$ value suggests a bimolecular rate-determining step of the hydrolysis reaction, preference being given to a nucleophilic reaction of water with the acyl group [76] because k_H/k_D isotope effects of similar magnitude are found as for the solvolysis of other activated acyl groups.

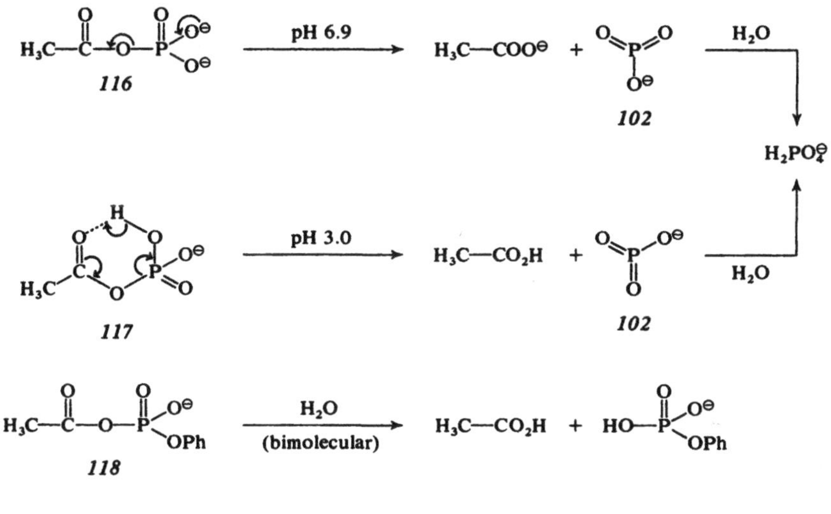

$\Delta S^{\neq}$ (39 °C) [eu]

116: + 3.7
117: − 3.6
118: −28.8

This mechanistic picture is supported by the following observations:

a) The isotope effect k_{H_2O}/k_{D_2O} for the hydrolysis of *118* is 2.5, but is insignificant for the hydrolysis of *116* and *117*.

b) The rate of solvolysis of *116* and *117* is unchanged or slightly increased in concentrated salt solutions while that of *118* is strongly decreased.

c) On addition of 30–50% of acetonitrile the rate of hydrolysis remains unchanged for *116*, is slightly reduced for *117*, and strongly reduced for *118*; points b) and c) suggest that a bimolecular reaction with H_2O is rate determining only for *118*.
d) Hydrolysis of acetyl phosphate at neutral pH and close thereto occurs with preferential P—O bond cleavage (>89% at pH 3.8 [77]).
e) Hydrolysis of the monoanion can already be easily observed at 39 °C while this is the case only at 100 °C with other monoester monoanions. This supports the particular significance assumed for the intramolecular proton transfer.

A situation similar to that in acetyl phosphate is also encountered in benzoyl phosphate [76]. Electron-attracting substituents on the phenyl ring accelerate the hydrolysis of the dianion (a linear relationship exists between log $k_{hydrol.}$ and the Hammett σ constants with $\varrho = 1.2$ and the linear log $k_{hydrol.}/pK_a$ relationship is the same as for the phosphoric monoaryl ester dianions [65]. On the other hand, hydrolysis of the monoanion is influenced only slightly by substituents in the phenyl ring. These observations can also be rationalized in terms of the decomposition mechanism to the $PO_3^{\ominus}$ ion formulated for *116* and *117*.

As for the acetyl phosphate monoanion, a metaphosphate mechanism has also been proposed [78] for the carbamoyl phosphate monoanion *119*. Once again, an intramolecular proton transfer to the carbonyl group is feasible. The dianion likewise decomposes in a unimolecular reaction but not with spontaneous formation of $PO_3^{\ominus}$ as does the acetyl phosphate dianion, but to $HPO_4^{2\ominus}$ and cyanic acid. Support for this mechanism comes from isotopic labeling proof of C—O bond cleavage and from the formation of carbamoyl azide in the presence of azide ions.

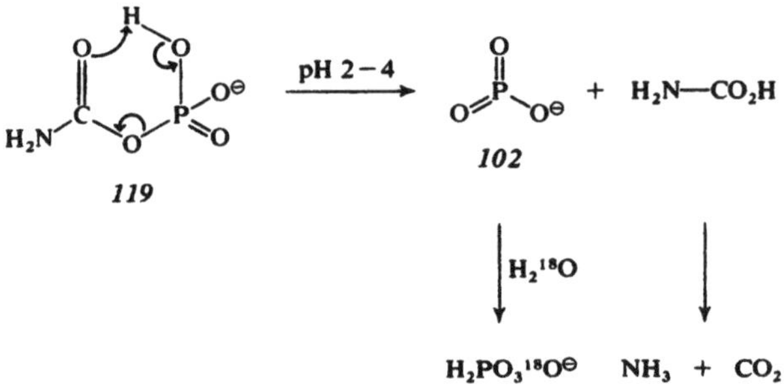

119

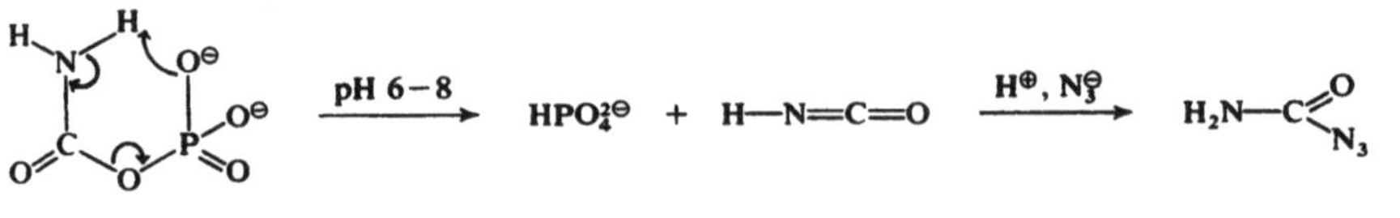

120

The mechanism of hydrolysis of o-carboxyaryl phosphates, whose dianions also hydrolyze much faster then, e.g., the phenyl phosphate monoanion [79,80] (maximum rate at about pH 4.8 and 25 °C [81]), was long a point of mechanistic contention. Thorough investigations [81] led to proposal of a fast initial transprotonation equilibrium according to *121*⇄*122*⇄*123* followed by slower decomposition of *123* to the salicylate ion and the $PO_3^{\ominus}$ ion *102*.

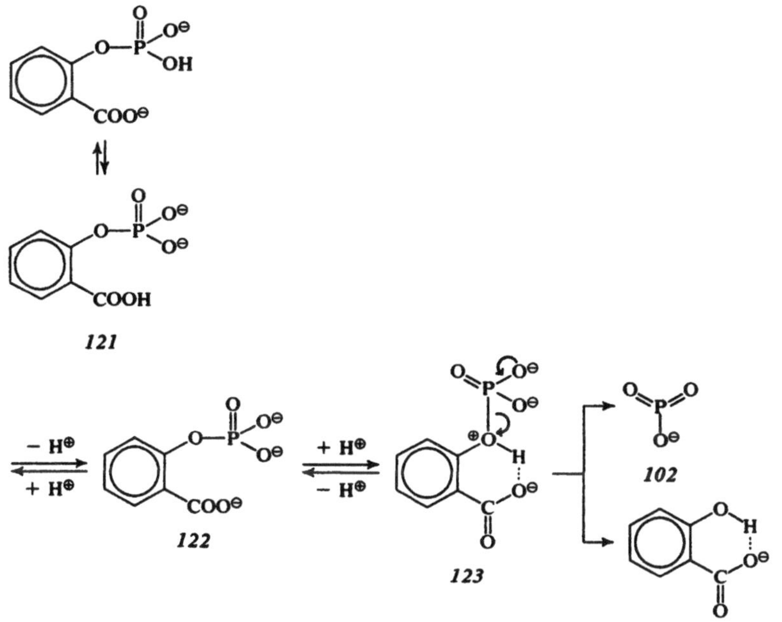

121

122

123

102

Hydrolysis of the mono- or dianion of the carboxylic acid *124a*, like that of the phosphoenol pyruvate monoanion *124b*, is formulated as proceeding via free metaphosphate ion *102* whose formation is also preceded by an intramolecular proton transfer according to *125*, which, however, does not involve the carbonyl function [82]. It should be mentioned, however, that the anion *124a* can apparently also undergo a competing hydrolysis mechanism formulated as proceeding via the cyclic oxyphosphorane *126* [83]. The latter can then afford the same products via the mixed anhydride *127*.

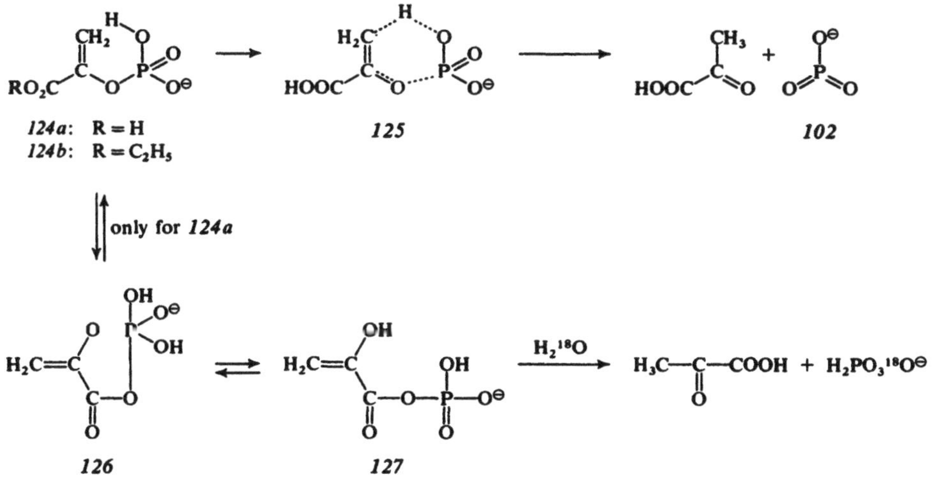

124a: R = H
124b: R = C_2H_5

125

102

126

127

As we have already seen, the monoanions and possibly also the dianions of phosphoric O-esters and also S-esters are potential metaphosphate precursors. Hydrolysis of the monoanions of phosphoramidates, however, will hardly involve a free metaphosphate in most cases. The transition state *128* of these reactions would appear to correspond to a "loosely coupled" S_N2 reaction in which bond breakage of the leaving group is further advanced than bond linkage to the attacking nucleophile [84], and which thus possesses considerable metaphosphate character. A metaphosphate reaction has hitherto been formulated for only a few individual cases, as for the hydrolysis of phosphorylurethane *129* [85] and phosphorylguanidine *130* [86]. The special reaction mechanism for *130* is underlined by finding that both the monoanion and the dianion, as well as the O-benzyl ester of *130*, are hydrolyzed at least 10^4 times slower than neutral *130*.

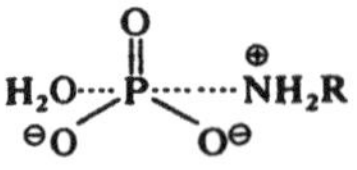
128

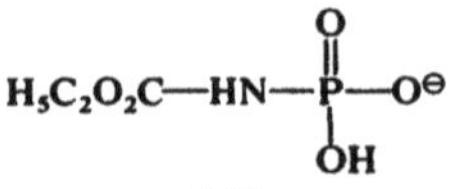
129

130

The possible mechanisms for solvolysis of phosphoric monoesters show that the pathway followed depends upon a variety of factors, such as substituents, solvent, pH value, presence of nucleophiles, etc. The possible occurrence of monomeric metaphosphate ion cannot therefore be generalized and frequently cannot be predicted. It must be established in each individual case by a sum of kinetic and thermodynamic arguments since the product pattern frequently fails to provide unequivocal evidence for its intermediacy. The question of how "free" the $PO_3^{\ominus}$ ion actually exists in solution generally remains unanswered. There are no hard boundaries between solvation by solvent, complex formation with very weak nucleophiles such as dioxane or possibly acetonitrile, existence in a transition state of a reaction, such as in *129*, or $S_N2(P)$ or oxyphosphorane mechanisms with suitable nucleophiles.

4.1.2 Hydrolysis of Diphosphates and Triphosphates

A mechanism proposed [87] for the alkaline hydrolysis of tetraethyl pyrophosphate, which is markedly accelerated by $HPO_4^{2\ominus}$ ions, has been substantiated by isotopic labeling [88]. The nucleophilic attack by $HPO_4^{2\ominus}$ on the symmetrical pyrophosphate *131* is considered to lead initially to the unsymmetrical P^1,P^1-diethyl pyrophosphate dianion *132* which decomposes spontaneously under the conditions of reaction to give the diethyl phosphate anion and $PO_3^{\ominus}$ *102*. The latter reacts with water to form "inorganic" phosphate and with alcohols such as methanol and ethylene glycol to produce alkyl phosphates.

Direct synthesis of P^1,P^1-diethyl pyrophosphate [89] was accomplished later. It is stable at pH 3, but hydrolyzes rapidly at pH 7/room temperature. Thus in this case it is the dianion which hydrolyzes faster than the monoanion, in contrast to the behavior of alkyl phosphates or pyrophosphate monoesters (see below).

It appears that $PO_3^{\ominus}$ can also be formed from pyrophosphate monoesters. The pH/rate profile of hydrolysis of γ-phenylpropyl pyrophosphate *133* again shows a maximum in the pH range of 3–6, i.e. for the monoanion. The rate of hydrolysis of *133* is about 2000 times greater than that of the symmetrical P,P'-di-γ-phenylpropyl

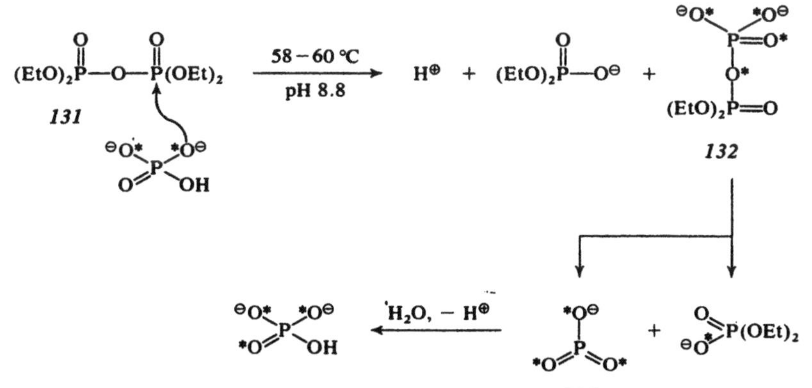

131 132 102

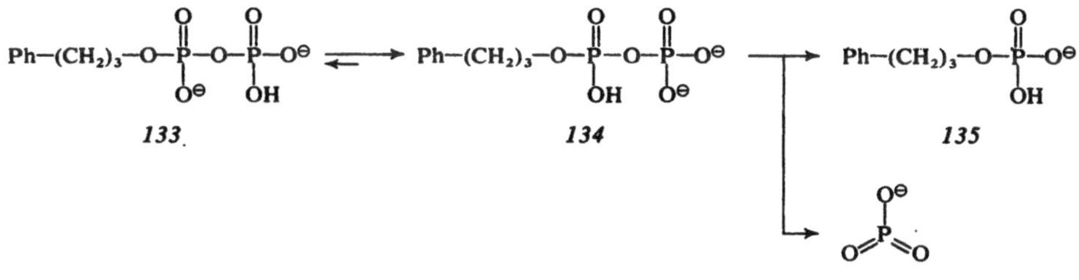

133 134 135 102

pyrophosphate, $ROPO_2$-O-$PO_2OR^{2\ominus}$, and is independent of nucleophilic or general base catalysis around pH 5. These facts are also regarded as evidence for the metaphosphate mechanism (*133*→*134*→*135* + *102*) [90]. Since the pH/rate profile of hydrolysis of γ-phenylpropyl diphosphate resembles that of adenosine diphosphate, and that of γ-phenylpropyl triphosphate in the region examined (pH 4–9) is similar to that of adenosine triphosphate, the metaphosphate mechanism must also be reckoned with in the hydrolysis of ADP and ATP in vitro.

4.1.3 Base Cleavage of β-Halophosphonic Acids

The β-bromophosphonic acid *136* is unstable in aqueous solution and decomposes slowly with formation of benzal acetophenone, hydrobromic acid, and phosphoric acid. Decomposition is instantaneous in alkaline solution. In contrast, the monophenyl ester of *136*, i.e. *137*, is stable and can be dissolved without decomposition in aqueous sodium carbonate [91].

β-Bromophosphonic acid *138*, α,β-dibromophosphonic acid *139*, and 2-bromovinyl-1-phosphonic acid *140* are hydrolyzed fast by aqueous sodium carbonate [92]. Compounds of structure comparable with that of *140* are the vinylphosphonic

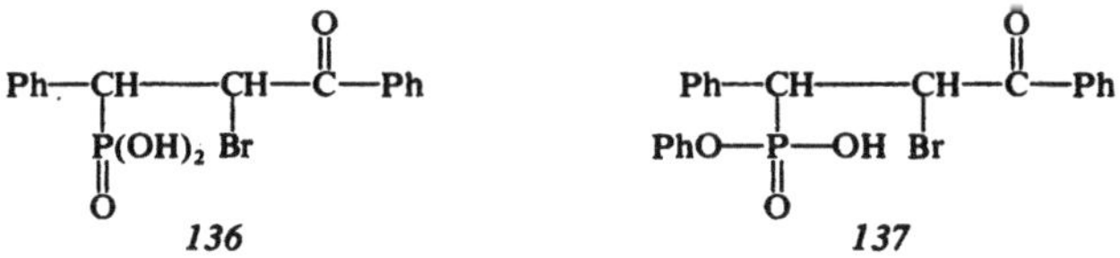

136 137

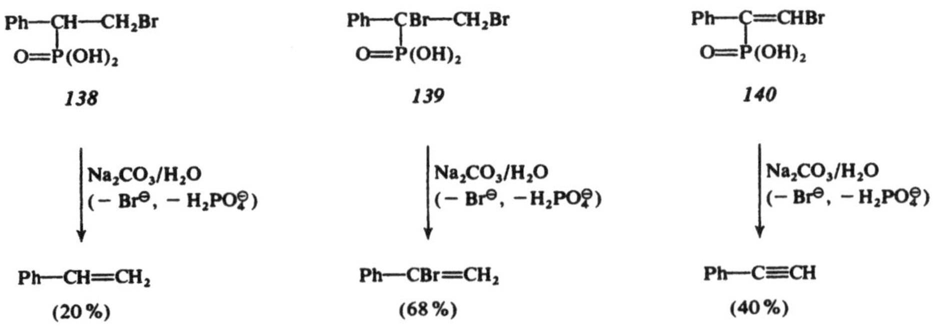

acids *141*. While *141a* (R = H) merely undergoes HCl cleavage with warm potassium hydroxide solution, the compound *141b* (R = OCH_3) which is unstable in pure form, undergoes elimination of $Cl^{\ominus}$ and the phosphonate group analogously to the case of *140* [93].

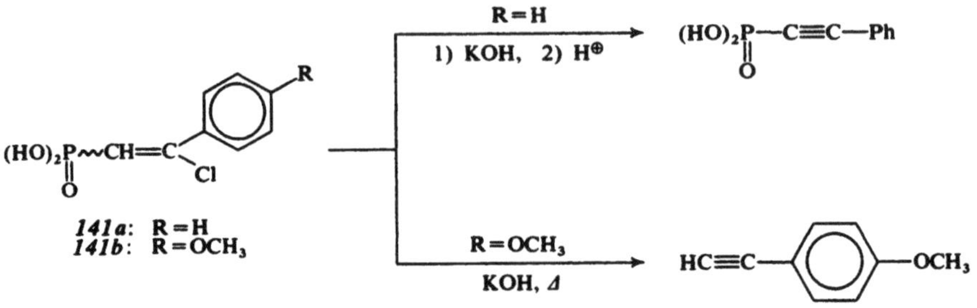

In a similar way, numerous β-chloroalkylphosphonic acids undergo fast and quantitative hydrolysis at pH > 5 [94]. A typical example is the fragmentation of 2-chlorodecyl-1-phosphonic acid in the presence of cyclohexylamine with formation of 1-decene (in contrast, other bases such as pyridine, triethylamine, and dimethylaniline effect only HCl elimination to form 1-decene-1-phosphonic acid). Added alcohols, such as ethanol, allyl alcohol, cyclohexanol, tert-butanol, and phenol are phosphorylated.

All these fragmentation reactions are best interpreted by a metaphosphate mechanism starting from the phosphonic dianion *142*, formulated as follows:

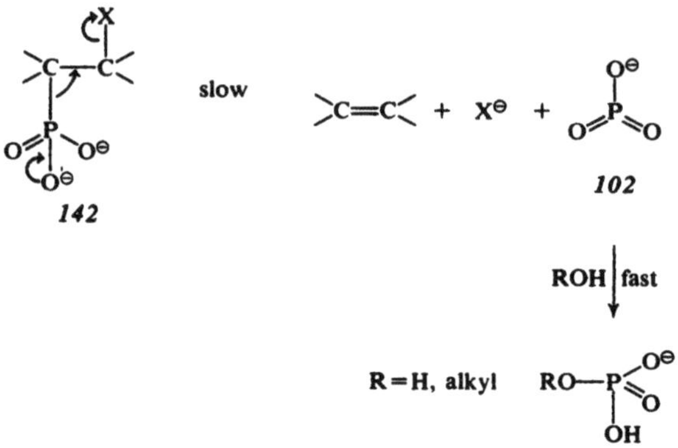

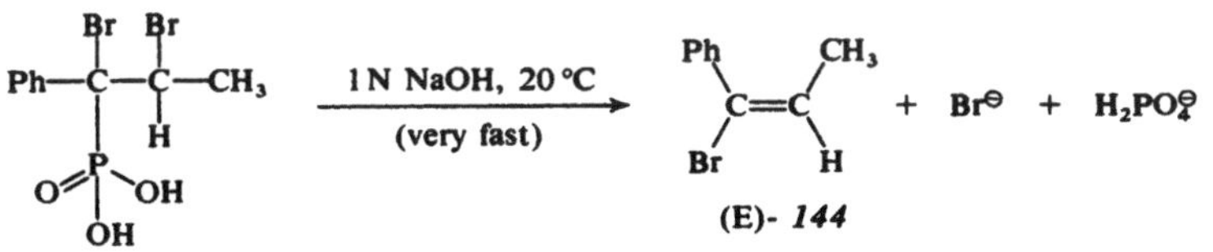

erythro- *143*

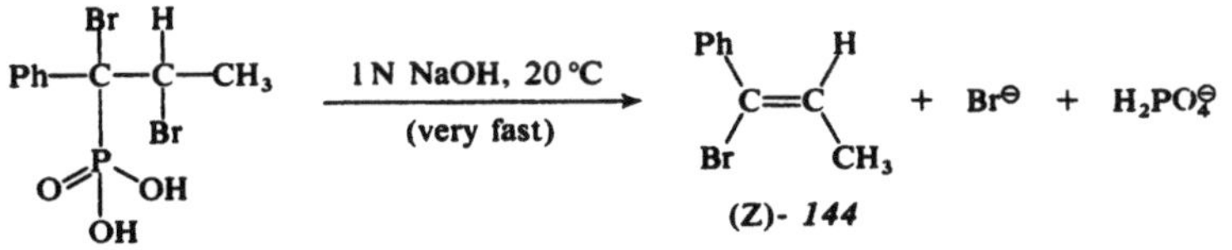

threo- *143*

Induction of elimination by nucleophilic attack of a solvent molecule on phosphorus ($S_N2(P)$) would represent a mechanistic alternative. However, since not only ethanol but also sterically demanding alcohols such as cyclohexanol and tert-butanol are phosphorylated on alcoholysis of 2-chlorodecyl-1-phosphonic acid, this mechanism appears unlikely. Neither alcohol is a suitable nucleophile.

It was shown that the fragmentation of *143* occurs stereospecifically in the sense of a trans-elimination: erythro-*143* gives the olefin (E)-*144* while threo-*143* furnishes (Z)-*144* [95].

Fragmentation of *143* is not limited to aqueous media. Reaction of threo-*143* with cyclohexylamine, triethylamine, or 2,6-lutidine in acetonitrile also leads to (Z)-*144* while the phosphorus appears in various polyphosphates, in the trimetaphosphate trianion (after 110 °C/5 h, before work-up), and in phosphoramidates [95].

4.1.4 Phosphorylation Reactions with the $PO_3^{\ominus}$ Ion

The unique character of the $PO_3^{\ominus}$ ion as an anionic Lewis acid is manifested in its great reactivity, which hitherto has precluded its isolation. Its pronounced electrophilic nature permits rapid phosphorylation of water, alcohols, amines, and even of phosphate ions in many of the solvolysis reactions described so far. This greatly pronounced reactivity relative to the isostructural $NO_3^{\ominus}$ ion demands an explanation. The reason does not lie in the weakness of the P-O bond, for its total overlapping population is greater than for the corresponding bond in $NO_3^{\ominus}$ owing to the participation of d orbitals [96 b)]. Calculation of the frontier orbitals in the two ions shows [96)], however, that the energy of the σ^* orbital is drastically lowered so that it becomes almost energetically comparable with the π^* orbital (see Fig. 3).

The orbital-controlled electrophilic reactivity of the $PO_3^{\ominus}$ ion can be understood as an interaction between an occupied donor orbital and the unoccupied σ^*- and π^*-orbitals of the $PO_3^{\ominus}$ ion after the manner of a 2-electron-3-orbital interaction. Under the influence of the attacking donor orbital, effective mixing of the energetically similar σ^*- and π^*-orbitals becomes possible, leading to a strongly bonding new MO of low energy. Owing to the unattainability of the high-energy σ^*-orbital, no such HOMO-LUMO interaction is possible in the $NO_3^{\ominus}$ ion. Furthermore, the extended electron density at the phosphorus will certainly also permit better interaction of the $PO_3^{\ominus}$ ion with nucleophiles than is the case in the $NO_3^{\ominus}$ ion.

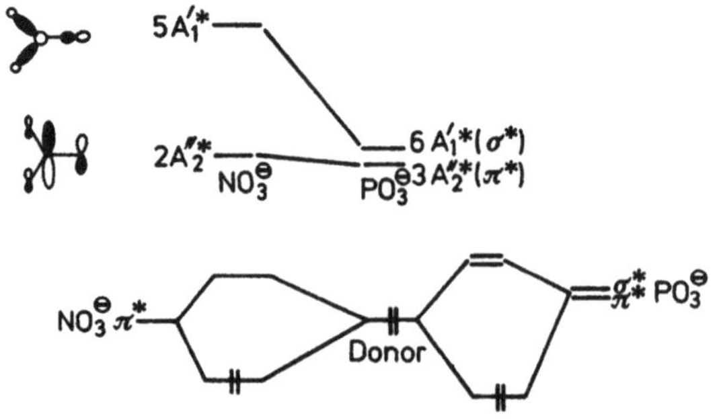

Fig. 3. Top: π* and σ* orbitals in $NO_3^{\ominus}$ and $PO_3^{\ominus}$ ions; Bottom: Interaction between occupied donor orbital and unoccupied acceptor orbitals of the $NO_3^{\ominus}$ and $PO_3^{\ominus}$ ions

The highly electrophilic character of the $PO_3^{\ominus}$ ion would suggest a very unselective phosphorylation behavior. For example, the ratio of alkyl phosphate to inorganic phosphate obtained in hydrolyses of phosphoric esters in water/alcohol mixtures should reflect the molar ratio of water and alcohol. This is indeed found in numerous cases, e.g. in the hydrolysis of phenyl and 4-nitrophenyl phosphate monoanions [97] or of 4-nitrophenyl phosphate dianions [65] at 100 °C in methanol/water mixtures of various compositions, as also in the solvolysis of the acetyl phosphate dianion at 37 °C [97] or of phosphoenol pyruvate monoanions [82]. Calculations of the free energy of the addition reactions of water and ethanol to the $PO_3^{\ominus}$ ion support the energetic similarity of the two reactions [98] (Table 4).

Wherever it is observed, nonselective phosphorylation has been viewed as an indication of the intermediacy of the metaphosphate anion. If phosphate ester hydrolysis should occur via an $S_N2(P)$ reaction, then a higher proportion of methyl phosphate than of inorganic phosphate would be expected, for example in CH_3OH/H_2O mixtures, since CH_3OH is to be regarded as a more effective nucleophile. In several cases, however, selectivity has been observed in the reactions of the (presumed) $PO_3^{\ominus}$ intermediate, thus disqualifying unselective phosphorylation as necessary evidence for a $PO_3^{\ominus}$ ion. For instance, methanol is more reactive than water towards the 2,4-dinitrophenyl phosphate dianion at 39 and 100 °C; the corresponding monoanion also "phosphorylates" more methanol than water at 39 °C, whereas it proves unselective towards a mixture of CH_3OH (30%)/H_2O at 100 °C. Examination of various alcohols shows that the 2,4-dinitrophenyl phosphate dianion phosphorylates up to 4 times more methanol, up to 2 times more ethanol, and less 1,1,1-trifluoroethanol

Table 4. Calculated free energies of addition and elimination of water and ethanol in the phosphate/metaphosphate system [98]

Reaction	ΔG° [kcal mol^{-1}]	log K
$H_2PO_4^{\ominus} \rightleftharpoons PO_3^{\ominus} + H_2O$	27 ± 2	−19.6 ± 1.8
$HPO_4^{2\ominus} \rightleftharpoons PO_3^{\ominus} + OH^{\ominus}$	36 ± 2	−26.4 ± 1.8
$EtOPO_3H^{\ominus} \rightleftharpoons PO_3^{\ominus} + EtOH$	25 ± 3	−18.1 ± 2.0
$EtOPO_3^{2\ominus} \rightleftharpoons PO_3^{\ominus} + EtO^{\ominus}$	37 ± 3	−27.0 ± 2.0

than water. 2-Propanol (94%) is not phosphorylated at all [65]. The formation of pyrophosphate on hydrolysis of acetyl phosphate in media of high salt concentration has also been viewed as evidence for the $PO_3^\ominus$ ion [76]. Hydrolysis of the acetyl phosphate *145* in 7.3 M $NaClO_4$ solution at pH 4.0, 5.6, and 9.7 leads to 9, 15, and 33% conversion, respectively, into pyrophosphate. The reduced availability of water thus apparently permits direct reaction of the $PO_3^\ominus$ ion with phosphate already formed (or possibly with acetyl phosphate still present).

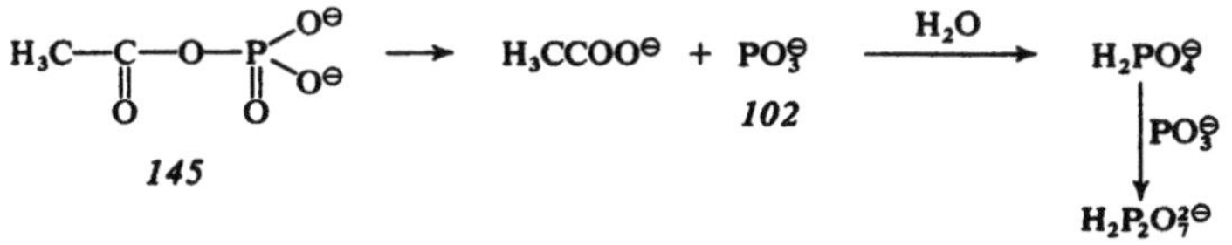

145 *102*

Less is known about unequivocal reactions of the $PO_3^\ominus$ ion with amines. This is partly because the phosphate esters examined undergo direct $S_N2(P)$ reaction with amines via oxyphosphoranes or can at least react in the boundary area between S_N1 (metaphosphate mechanism) and $S_N2(P)$ [37] or because the reaction actually occurs at another part of the molecule (simple primary and secondary amines react with 2,4-dinitrophenylphosphate to give 2,4-dinitroanilines [99]).

Evidence for the phosphorylation of cyclohexylamine with the $PO_3^\ominus$ ion can be deduced from the dehydrohalogenation of α-chlorodecylphosphonic acid (see also Sect. 4.1.3); the expected cyclohexylphosphoramidate could not, however, be obtained pure [94].

An impressive example of phosphorylation of an amine from potential metaphosphate precursors is provided by the following three-phase experiment [100]: An insoluble polymer P_1, carrying the metaphosphate precursors acyl phosphate or β-haloalkyl phosphonate as functional groups, and a polymer P_2 with a primary amine as functional group are suspended in a solvent. Since a direct reaction between the functionals groups bound to the polymers is impossible, the observed phosphorylation of the amine must be due to a reactive intermediate transported through the solution. Whether it is the monomeric $PO_3^\ominus$ ion or perhaps a $PO_3^\ominus$/solvent complex (e.g. $PO_3^\ominus \cdot$ dioxane analogous to the known $SO_3 \cdot$ dioxane complex), cannot be decided by experiment. Only the analogy with the phosphorylation behavior of acyl phosphates (see Sect. 4.1.1) or β-haloalkyl phosphonates (see Sect. 4.1.3) in

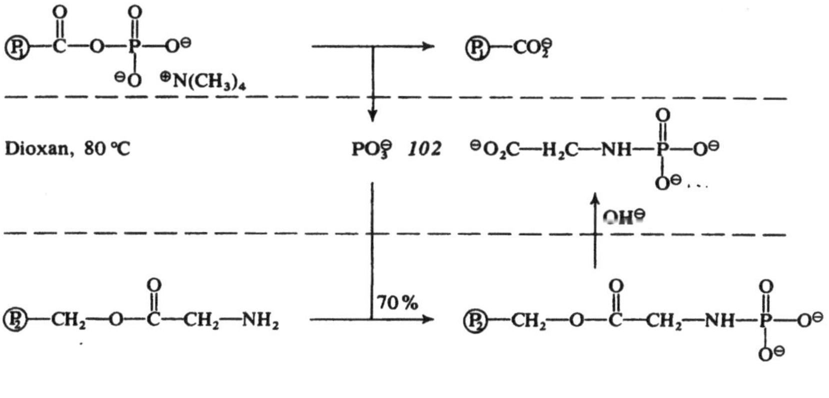

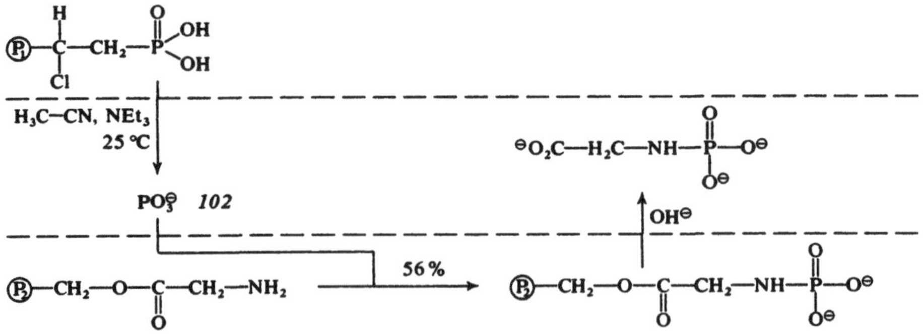

solution and the additional arguments presented there for a metaphosphate mechanism make a $PO_3^{\ominus}$ intermediate appear plausible. A pyrophosphate can also be ruled out as intermediate since it does not phosphorylate P_2-amine under the conditions of the reaction.

4.2 Alkoxydioxophosphoranes

Esters of monomeric metaphosphoric acid like *147* are just as unisolable as metaphosphoric acid itself or its anion. All the metaphosphoric esters described in the previous literature [101] are probably actually mixtures of oligomeric metaphosphoric esters, or well defined compounds such as cyclic trimetaphosphate.

Thus the reaction product of diethyl ether and diphosphorus pentoxide, a pale yellowish syrupy mass, was designated as ethyl metaphosphate (*147*), other ethoxy-containing compounds give the same product [102, 103]; it can be reprecipitated from chloroform/ether and reacts with aniline to give the phosphoramidate *148* [104] which is conceivably a trapping product of ethyl metaphosphate *147*, or a cleavage product of polymeric metaphosphates.

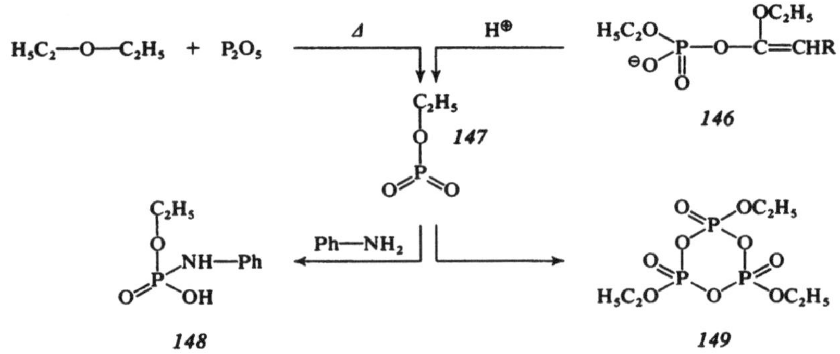

It was demonstrated that the IR spectrum of the "ethyl metaphosphate" *147* accessible from the enol phosphate *146* is identical with that of the cyclic triethyl metaphosphate *149* [105] synthesized by an independent route.

Finally, the product resulting on reaction of benzoic anhydride with diphosphorus pentoxide was also regarded as monomeric benzoyl metaphosphate. The description of this product as a "glassy translucent solid" raises doubts as to its monomeric character. Metaphosphates have also been postulated as reactive intermediates in the acylation of thiophene with acyl halides or carboxylic anhydrides in the presence of catalytic amounts of diphosphorus pentoxide [106].

More recent studies have shown that monomeric metaphosphates such as *147* are just as unisolable as the metaphosphate ion *102*, and are even more electrophilic. Generation of metaphosphates is accomplished mainly in two ways, i.e. by thermal or photochemical fragmentation reactions, on the one hand, and by decomposition of suitably activated phosphates on the other.

4.2.1 Thermal and Photochemical Fragmentation

Thermolytic processes are so far known to play only a minor role in the generation of the $PO_3^{\ominus}$ ion. An isolated case is the thermolysis of the monosodium salt of acetonylphosphonic acid at 150 °C, which leads to acetone and sodium polymetaphosphate [107]. A cyclic fragmentation mechanism, as known for β-ketocarboxylic acids, could lead to $PO_3^{\ominus}$ in this case.

On the other hand, numerous examples are already known in which monomeric metaphosphoric esters are generated by thermolysis reactions. Most worthy of mention in this context is the gas phase pyrolysis of the cyclic phosphonate *150* which leads via a retro-Diels-Alder reaction to butadiene and monomeric methyl metaphosphate (*151*) [108, 109, 110]. While most of the phosphorus appears as pyrophosphate and trimeric and polymeric metaphosphate, a low percentage (<5%) of products *152* and *153* is also found on condensation of the pyrolyzate in a cold trap containing diethylaniline or N,N,N',N'-tetraethyl-m-phenylene-diamine. The

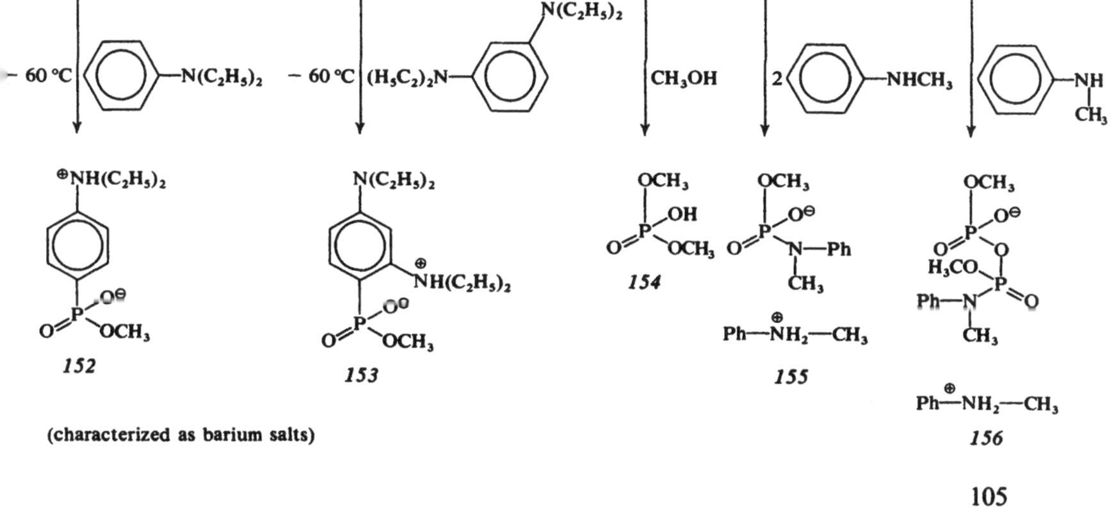

electrophilicity of *151* is apparently so great that electrophilic substitution of electron-rich aromatic compounds can occur even at —60 °C.

The reaction of *151* with methanol to give dimethyl phosphate (*154*) or with N-methylaniline to form the phosphoramidate *155* and (presumably) the pyrophosphate *156* complies with expectations. The formation of dimethyl phosphate does not constitute, however, reliable evidence for the formation of intermediate *151* since methanol can also react with polymeric metaphosphates to give dimethyl phosphate. On the other hand, reaction of polyphosphates with N-methylaniline to give *156* can be ruled out (control experiments). The formation of *156* might encourage speculations whether the reaction with N,N-diethylaniline might involve initial preferential reaction of monomeric methyl metaphosphate via interaction with the nitrogen lone pair to form a phosphoric ester amide which is cleaved to phosphates or pyrophosphates on subsequent work-up (water, methanol). Such a reaction route would at least explain the low extent of electrophilic aromatic substitution by methyl metaphosphate.

A further indication of formation of monomeric metaphosphoric esters or monomeric metaphosphoric acid itself by gas phase fragmentation comes from the behavior of 3-oxo-2-butyl phosphates in the mass spectrometer [111].

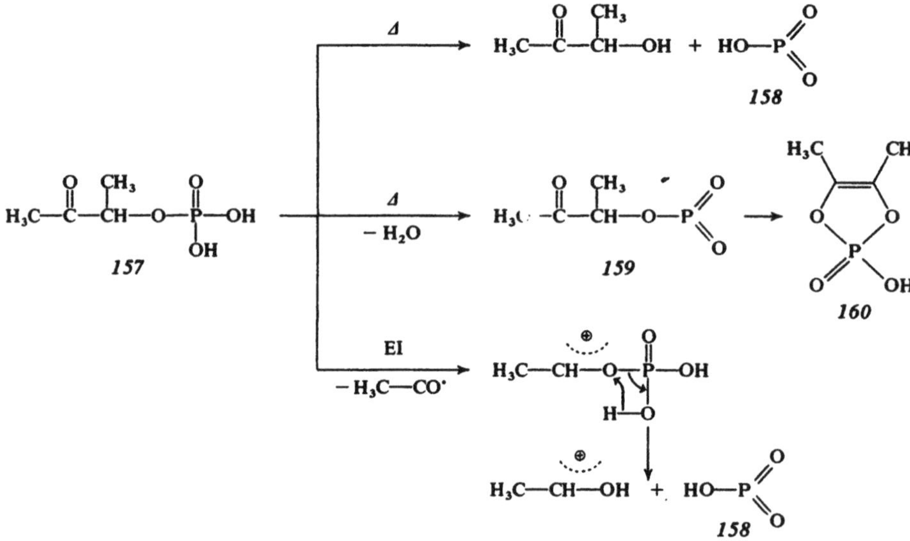

157 *158* *159* *160* *158*

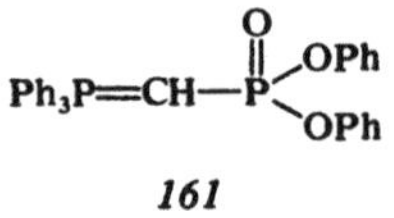

161

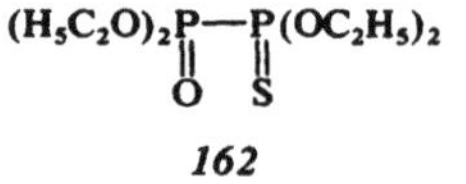

162

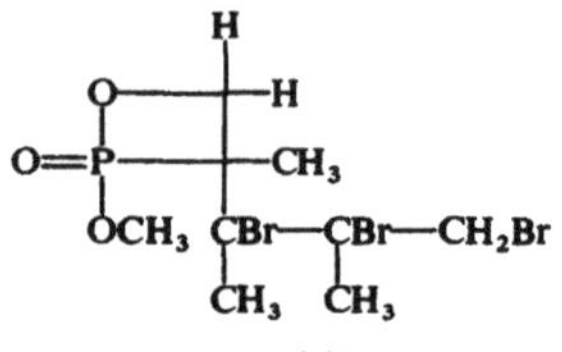

163

Thus, for example, *157* decomposes thermally (ca. 200 °C) into acetoin and monomeric metaphosphoric acid (*158*); both fragments appear in the mass spectrum.

Thermal dehydration of *157* to acetoin metaphosphate *159* and its rapid isomerization to the cyclic enediol phosphate *160* has not been confirmed in all details but is nevertheless in accord with all the data available. Under EI conditions, *157* yields a fragment $C_2H_6O_4P^{\oplus}$ which decomposes to give $C_2H_5O^{\oplus}$ and once again metaphosphoric acid (*158*). The monomethyl ester of *157* and its monoanion also undergo analogous fragmentation [112].

Mass-spectrometric fragmentation patterns suggesting elimination of alkyl and aryl metaphosphates have also been observed with compounds *161–163* [113–115]. However, the peak expected for metaphosphate itself is actually observed only in the case of *162*.

The λ^5-phosphorin *164* and the bicyclic compound *165* are precursors of isopropyl metaphosphate [116]. Thermal fragmentation of *165* leads via [2 + 2]cycloreversion to triphenyltoluene *166* and isopropyl metaphosphate *167*. The latter is identified as isopropyl phosphate after reaction with water. The mass spectrum of *165* is also dominated by this fragmentation picture (m/e 442 (16%) = M^+; m/e 320 (100%) = M^+ − *167*).

Photochemical fragmentation *165*→*166*+*167* is also feasible. Isopropyl phosphate and isopropyl methyl phosphate are expectedly found after reaction of the photolysis solution with water or methanol. It would appear that *166* (and possibly also *167*) can arise directly from *164* and not only by photochemical cycloreversion of *165* since *166* is formed together with *165* even at wavelengths in the range where *165* is known to be stable.

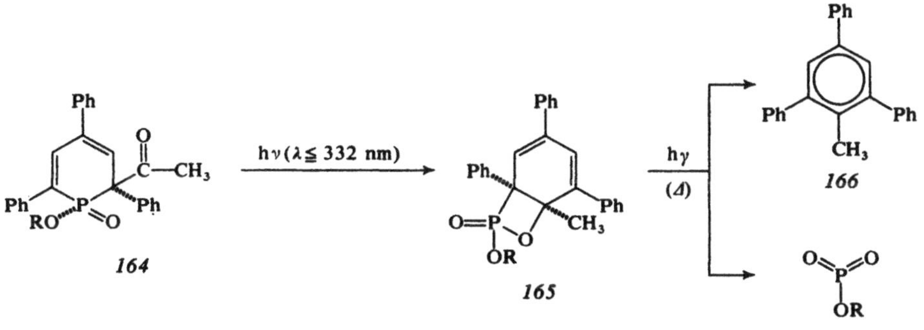

R = CH(CH₃)₂

4.2.2 Elimination Reactions of Activated Phosphates and Phosphonates

By analogy with the formation of the $PO_3^{\ominus}$ ion from phosphoric monoester anions, it might be expected that phosphoric diester anions fragment to give monomeric metaphosphates if one of the ester functions contains a good leaving group. This is not the case, for example, in dialkyl phosphates. They are hydrolyzed sufficiently fast only in strongly acidic media, with preferential C—O bond cleavage, as demonstrated by hydrolysis in $H_2{}^{18}O$.

The characteristic features of hydrolysis of diaryl phosphate monoanions (pronounced influence of leaving group on the rate of hydrolysis, $k_{H_2O}/k_{D_2O} = 1.6$, and $\Delta S^{\neq} = -25$ eu [117]) also fail to support a metaphosphate mechanism [118]. Hydrolysis of the acetyl phenyl phosphate monoanion is likewise far slower than that of the acetyl phosphate monoanion or dianion which both fragment to the $PO_3^\ominus$ ion (see also Sect. 4.1.1) [76]. Hydrolysis of the phenyl acetyl phosphate anion takes place preferentially in a bimolecular reaction with C—O bond cleavage. It has recently been shown, however, that the metaphosphate mechanism is also possible to a small extent [100b]; under more drastic reaction conditions polymer-bound acylphosphate ester *168* transfers 8% of its phenyl metaphosphate structural moiety *169* to amino acetic ester bound to a second polymer. The role played by proton sponge in the generation or possible stabilization of *169* is unclear.

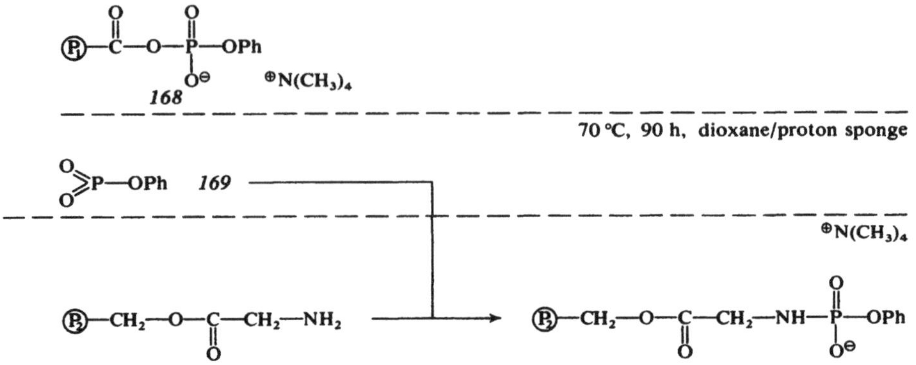

Another instructive example is provided by a series of α-phenyl-α,β-dibromophosphonates *170*, *171*, *172*. While the phosphonate dianion *170* fragments instantaneously at room temperature with formation of the $PO_3^\ominus$ ion (see also Sect. 4.1.3), the analogous reaction of the phosphonic monoester anion *171* leading to methyl metaphosphate *151* requires more drastic conditions and is at least 1000 times slower; the diester *172* is essentially stable under the reaction conditions described for *171*; addition of triethylamine leads to slow demethylation [110]. The behavior of *171* contrasts with that of "simple" β-haloalkylphosphonic monoesters which merely eliminate HHal on treatment with bases [94]. Thus it is the possibility of formation of a phenyl-conjugated double bond which supports the fragmentation of *171* to olefin + *151*.

The formation of *151* from the phosphonate *171* could be proved only by indirect means. Electron-rich aromatic compounds such as N,N-diethylaniline and N,N,N′,N′-tetraethyl-m-phenylenediamine [110,119] and N-methylaniline [120] are phosphorylated in the para- and in the ortho- plus para-positions by *151*. Furthermore, *151* also adds to the nitrogen lone pair of aniline to form the corresponding phosphoramidate. Considerable competition between nucleophiles of various strengths for the monomeric methyl metaphosphate *151* — e.g. aromatic substitution of N,N-diethylaniline and reaction with methanol or aromatic substitution and reaction with the nitrogen lone pair in N-methylaniline — again underline its extraordinary non-selectivity.

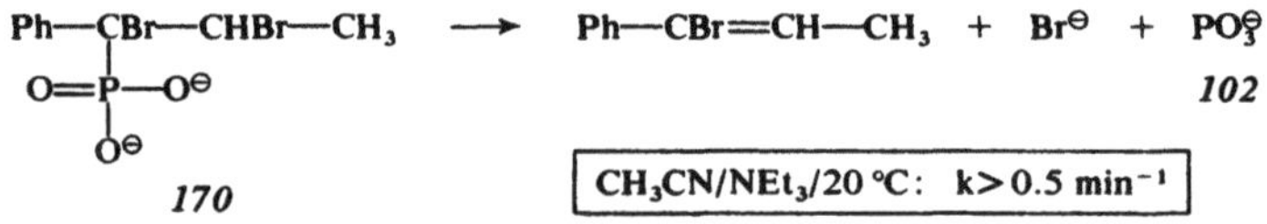

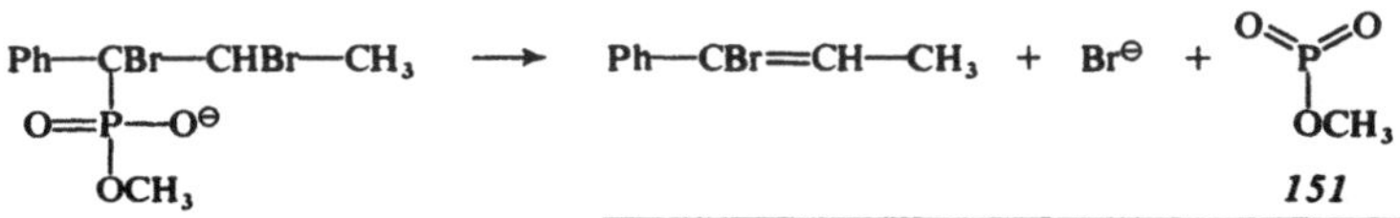

171 *151*

$CH_3CN/NEt_3/70\,°C$:	$k = 0.022\ min^{-1}$ for erythro-*171*
	$k = 0.032\ min^{-1}$ for threo-*171*

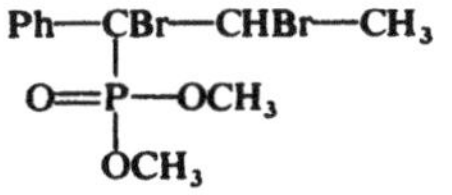

172

$CH_3CN/70\,°C/1\,h$:	< 5% conversion
$CH_3CN/NEt_3/1\,h$:	~10% demethylation

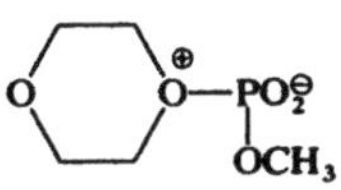

173

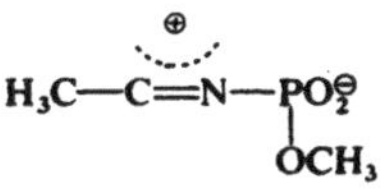

174

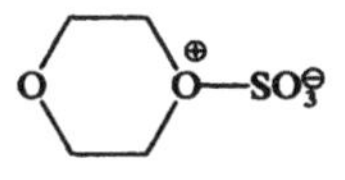

175

The extent to which *151* phosphorylates the aromatic amine in the phenyl ring is highly dependent upon the solvent. For instance, aromatic substitution of N-methylaniline is largely suppressed in the presence of dioxane or acetonitrile while phosphoramidate formation shows a pronounced concomitant increase. The presence of a fourfold excess (v/v) or pyridine, acetonitrile, dioxane, or 1,2-dimethoxyethane likewise suppresses aromatic substitution of N,N-diethylaniline below the detection limit. It appears reasonable to assume that *151* forms complexes of type *173* and *174* with these solvents — resembling the stable dioxane-SO_3 adduct *175* — which in turn represent phosphorylating reagents. They are, however, weaker than monomeric metaphosphate *151* and can only react with strong nucleophiles.

If monomeric *151* is generated from the phosphonic monoester *171* (OH in place of $O^{\ominus}$) in the presence of 2,2,6,6-tetramethylpyridine as base, then it also adds to carbonyl compounds [119,120]. Thus acetophenone is smoothly phosphorylated to the corresponding enol phosphate in 90% yield.

Another interesting topic is the occurrence of monomeric metaphosphate in oligonucleotide synthesis and its function as an active phosphorylation reagent. Among the chemical methods of oligonucleotide synthesis, the reaction of nucleosides with nucleoside phosphates in the presence of activating reagents such as N,N-dicyclohexylcarbodiimide, arylsulfonyl chlorides, or phosphoric halides occupies an important place [121]. An appropriate example is the linkage of 5′-O-tritylcytidine (*176*) (with protected amino group) with 3′-O-acetylthymidine 5′-phosphate (*177*) in the presence of N,N-dicyclohexylcarbodiimide (DCC), which furnishes *178*.

The intermediacy of a nucleoside metaphosphate in this synthesis was first discussed by Todd [122]. It could arise from the anhydrides *179a–c* formed as primary products and should effect phosphorylation of the nucleoside component.

^{31}P-NMR spectroscopic studies on the reaction course of the dinucleotide synthesis from 3′-O-acetylthymidine 5′-phosphate (pT-Ac) (*180a*) and 5′-O-tritylthymidine (Tr-T) in the presence of triisopropylbenzenesulfonyl chloride (TPS) confirm the metaphosphate hypothesis [123,124]. Successive addition of 0.5 equiv.

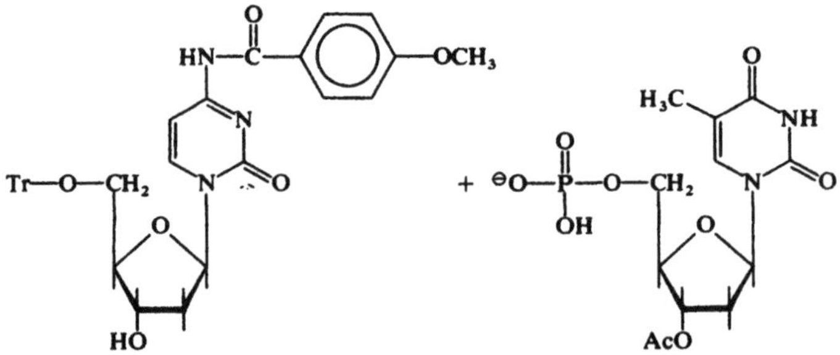

176 *177*

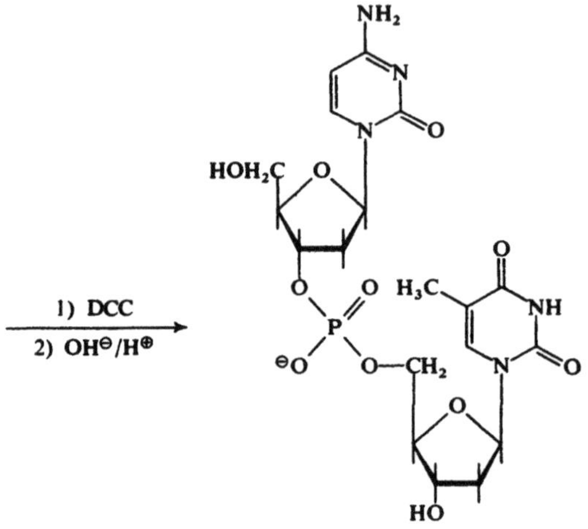

178

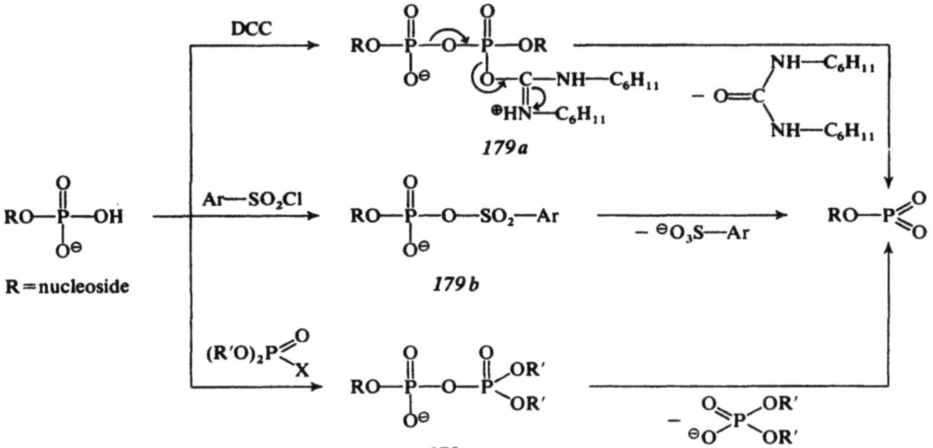

179a *179b* *179c*

portions of TPS to a pT-Ac solution triphosphate *182a*, and finally to 3'-O-acetylthymidin-5'-yl metaphosphate (*183a*). The cyclic trimetaphosphate *184a*, likewise conceivable as an intermediate and phosphorylation reagent [125], is not formed. Completely analogous behavior is observed with p-nitrophenyl phosphate *180b*, which gives p-nitrophenyl metaphosphate *183b*.

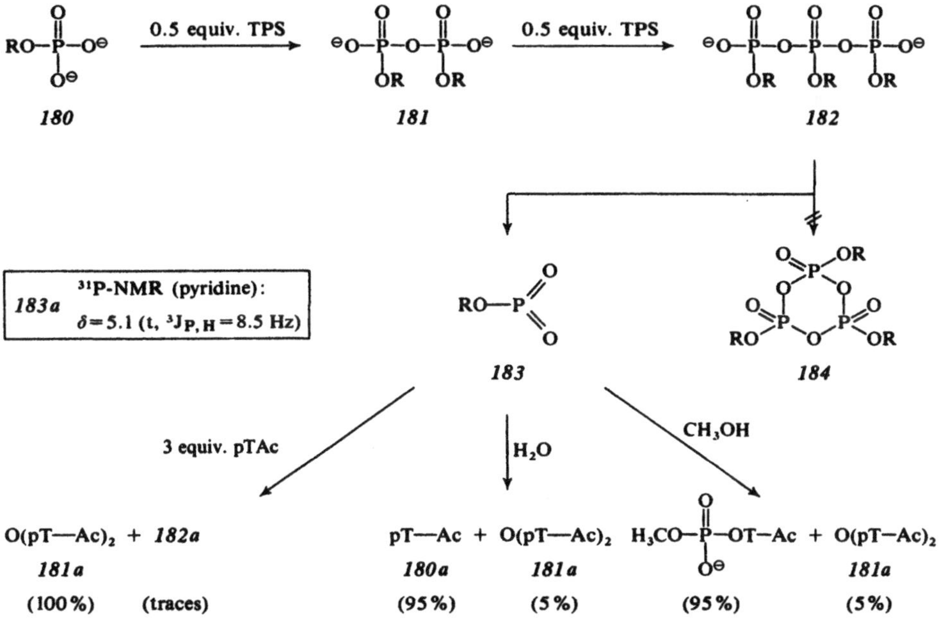

180a–184a: R=

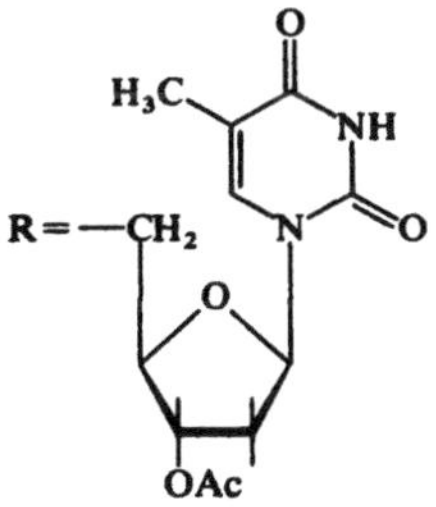

180b–184b: R=

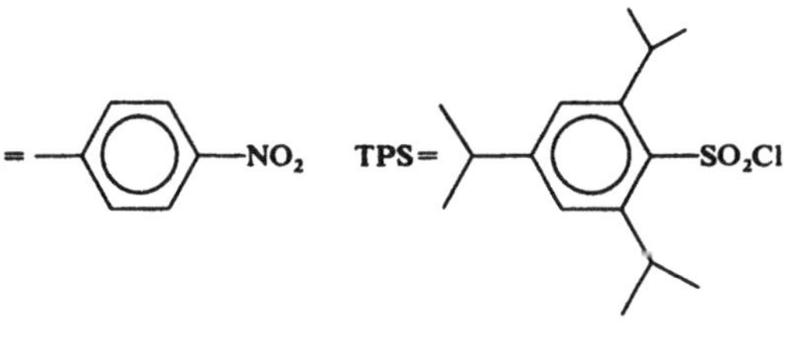

The reaction of the metaphosphate *183a* with alcohols and water is in accord with expectation; direct phosphorylation of water or methanol predominates with formation of the phosphate *180a* or the corresponding monoethyl phosphate. The "second phosphorylation" of pT-Ac *180a* already formed by metaphosphate still present becomes an insignificant side reaction under these conditions. Reaction with 5'-O-tritylthymidine (*185*) proceeds slowly to give the phosphoric diester *186*, and thence in a fast process to the pyrophosphate triester *187* which is also a phosphorylating agent.

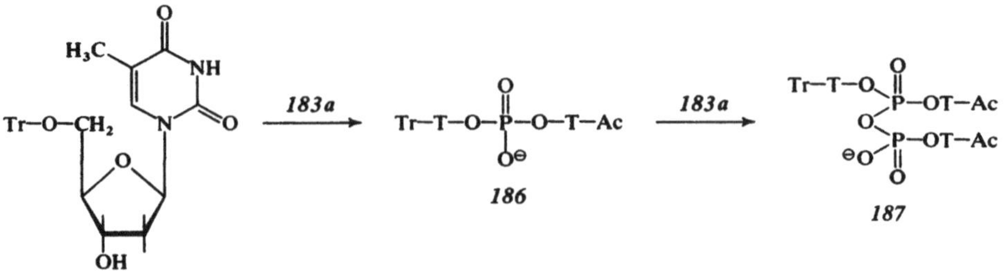

(Tr—T) *185*

186

187

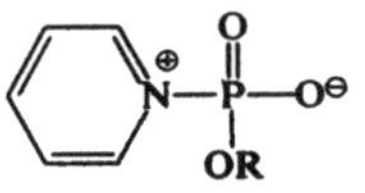

188

These facts provide a basis for setting up a mechanistic scheme for oligonucleotide synthesis with arylsulfonyl chlorides as activating reagents, as can be formulated in the simplest case with participation of nucleoside metaphosphate [123]. Formation of the pyrophosphate *181* and the triphosphate *182* accordingly proceeds by reaction of the metaphosphate *183* with the starting nucleotide *180* or with pyrophosphate *181*, respectively. As a result of the strong nucleophilic nature of the latter, the metaphosphate is immediately scavenged in the first stage of the reaction and does not therefore accumulate. The reaction of *181* or *182* with the arylsulfonyl chloride leads to adducts comparable with *179b* which can also be regarded as metaphosphate source within this scheme.

The fact that the ^{31}P-NMR signal of *183a* can only be observed in pyridine-containing solution provides food for thought [124]. Viewed in conjugation with the idea that alkyl metaphosphates could form adducts such as *173* and *174* [119,120] as discussed above, formulation as a zwitterionic pyridine/metaphosphate adduct (*188*) seems reasonable. Similar adducts have also been found in the reaction of TPS with dinucleotides and trinucleoside diphosphate [126]. In any case, the reactions of *183* or *188* are in full accord with the expected properties of a monomeric metaphosphate and its reactivity towards alcohols is far greater than that of all other reactive phosphorylation intermediates which can arise on reaction of TPS with oligonucleotides [126].

On use of N,N'-dicyclohexylcarbodiimide instead of sulfonyl chlorides as condensation reagent in oligonucleotide synthesis, then the pyro-, tri- and tetraphosphate stages are again involved [124]. The metaphosphate *183a* is found in small amounts by ^{31}P-NMR spectroscopy, but again no cyclic trimetaphosphate *184* can be detected, which would also be a possible phosphorylation reagent.

It is still unclear how much general validity attaches to the metaphosphate mechanism as formulated for the present variant for the oligonucleotide synthesis. Depending upon the reaction conditions, preparative method, activating reagent, and length of the nucleotide block, other mechanisms may dominate.

A hydroxymethylpolystyrene charged with ca. 17% phosphate groups, for example, is unable to phosphorylate alcohols in the presence of mesitylene-sulfonyl chloride [127]. On the other hand, phosphoric diesters are formed when OH-containing polymers are allowed to react with phosphoric monoesters under the same conditions. Since the phosphate-containing polymer contains ca. 60% isolated phosphate groups and ca. 30% pyrophosphate groups, it was concluded that phosphorylation does not proceed via monomeric or dimeric phosphates but instead via higher oligomers, for example, via the cyclic trimetaphosphate *184* which is activated for phosphorylation by sulfochloride [127,128].

4.3 Alkoxy(Amino)-iminooxophosphoranes and Alkoxy-iminothiophosphoranes

As already mentioned, three-coordinate P(V) compounds are obviously extremely short-lived and reactive intermediates as long as they contain a P=O or P=S bond. According to our present knowledge, this also applies to nitrogen-substituted oxo- and thio-phosphoranes *189–191*. Their intermediacy, especially in hydrolysis reactions of suitable (thio-)phosphoric amides is not undisputed. Compounds of this type having at least one hydrogen atom attached to nitrogen hydrolyze much faster in alkaline media than in neutral, provided that the phosphorus bears a suitable leaving group. Appropriate examples are alkylaminophosphoric chlorides *192a–c* [129,130,131], alkylaminophosphoric fluorides [132], 4-nitrophenyl dianilinophosphate *193* [133], and thiophosphoric amide chlorides such as *194* [134]. The hydrolysis of *192b* catalyzed by 2,6-lutidine should also be mentioned in this connection [130]. A "metaphosphorimidate" mechanism has been formulated to account for these findings [129,130,135].

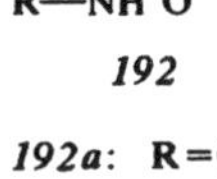
189

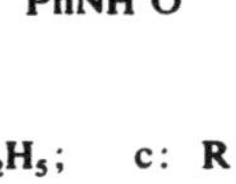
190

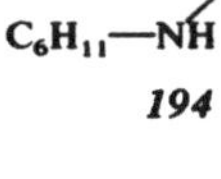
191

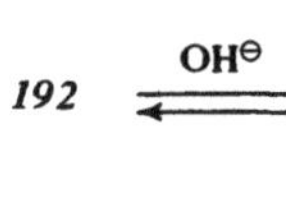
192

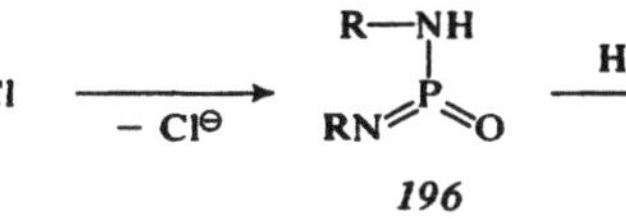
193

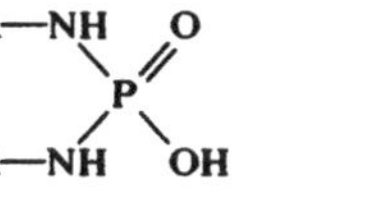
194

192a: $R=CH_3$; b: $R=C_2H_5$; c: $R=C_3H_7$

195 *196* *197*

According to this scheme, *192* first affords the anion *195* in an initial protonation/deprotonation equilibrium; *195* then undergoes rate-determining unimolecular decomposition to the metaphosphorimidate *197*, which reacts fast with water to give the phosphoric acid *197*.

Recent kinetic studies on thiophosphoric aryl ester dianilides suggest analogous decomposition. The rate law observed is in agreement with a hydrolysis scheme in which both the monoanion and the dianion decompose to metaphosphorothioimidate and its anion, respectively, which then react fast with water [133].

The following considerations and observations are in accord with the metaphosphorimidate solvolysis mechanism *192*→*195*→*196*→*197*:

a) If the phosphoric acid *197* were formed by direct $S_N2(P)$ reaction from the anion *195*, then we would expect *195* to be attacked slower by a nucleophile (even by a neutral molecule like H_2O) than the neutral molecule *192*. The opposite is actually the case.

b) The necessity of the presence of one or several NH functions for fast hydrolysis is impressively demonstrated by comparison with aminophosphoric acid derivatives whose N atoms are completely substituted. Thus *192c* is hydrolyzed in alkaline medium 4×10^6 times as fast as bis(dimethylamino)phosphoryl chloride [131]; and the 2nd order rate constant of alkaline hydrolysis of *193* is ca. 2×10^4 times higher than that of 4-nitrophenyl phosphorodimorpholidate which is fully substituted at the N-atoms [133].

   Steric reasons for differences between singly and doubly substituted amino derivatives can presumably be ruled out since it has been shown that nucleophiles other than $OH^{\ominus}$, such as $F^{\ominus}$, $N_3^{\ominus}$, and pyridine, react comparably fast with *192c* and bis(dimethylamino)phosphoryl chloride [131].

c) Since metaphosphorimidates like *196* are planar, attack by nucleophile should occur with equal probability from either side, and enantiomeric derivatives should be hydrolyzed with racemization. Very few studies have so far been devoted to this question. It has been found, however, that optically active N-cyclohexyl phosphoric methyl ester chloride (*198*) undergoes almost complete racemization on alkaline hydrolysis (dimethoxymethane/water, 1:1) while neutral hydrolysis proceeds with retention of optical activity (presumably stereospecifically) in accord with the $S_N2(P)$ or addition/elimination mechanism assumed for the latter reaction [134].

On the other hand, it is found that only partial racemization occurs on alkaline hydrolysis of optical active *198* in aqueous methanol [136] and no racemization takes place in the hydrolysis of *199* in dioxane/water [137]. Moreover, the latter reaction is only ca. 80 times faster at 29 °C than that of the analogous morpholide *200*, for which a metaphosphorimidate mechanism is precluded a priori by the absence of an NH function and whose hydrolysis is likewise stereospecific [137]. Clearly a free metaphosphorimidothioate of type *191* cannot be involved in this case. The experimental findings are compatible, however, with the hypothesis that the nucleophile water attacks a metaphosphorothioimidate/phenolate associate *201*. The question of how free metaphosphates occur in solution is of a general nature: it has also been considered in the previous Section.

Attempts to trap the metaphosphorimidate *196* (R = n-C_3H_7), which represents

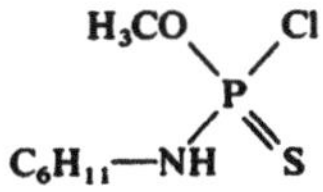

198

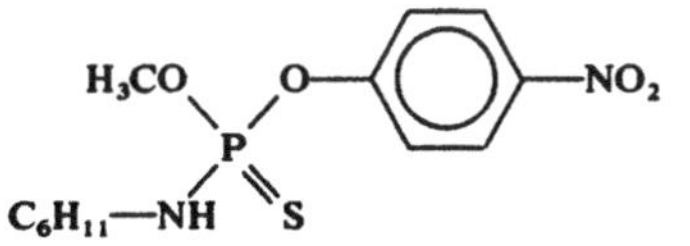

199

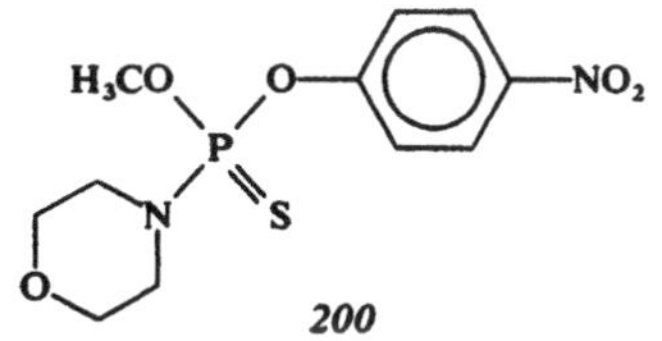

200

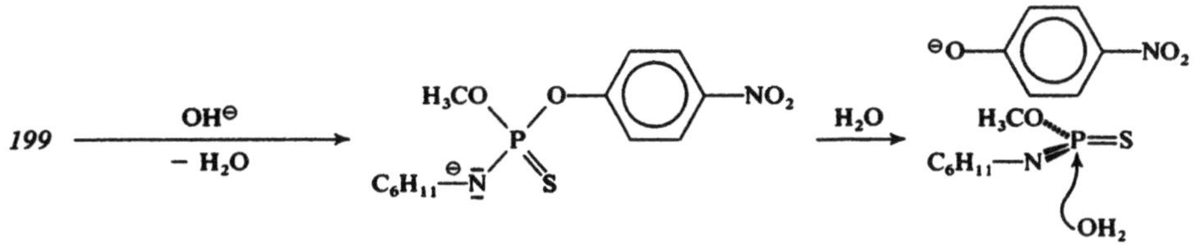

201

the presumed intermediate of alkaline hydrolysis of *192c*, with nucleophiles such as $F^{\ominus}$, $N_3^{\ominus}$ or imidazole were unsuccessful [131].

Powerful evidence for the intermediacy of a metaphosphorimidate of type *189* nevertheless comes from the product spectrum obtained on reaction of *202* with KOH in ethanol or propanol. The principal products dimethyl sulfide and potassium O-ethyl (or propyl) phosphoramidate can be readily rationalized in terms of the reaction sequence presented in the Scheme [138].

It should be mentioned, however, that the phosphoramidothioate *202* can undergo hydrolysis by another mechanism which becomes operative above all in polar solvents (e.g. aqueous KOH, and less so in methanol or acetone). P—O bond cleavage occurs, presumably via an addition/elimination mechanism, while the metaphosphorimidate pathway is characterized by P—S bond cleavage.

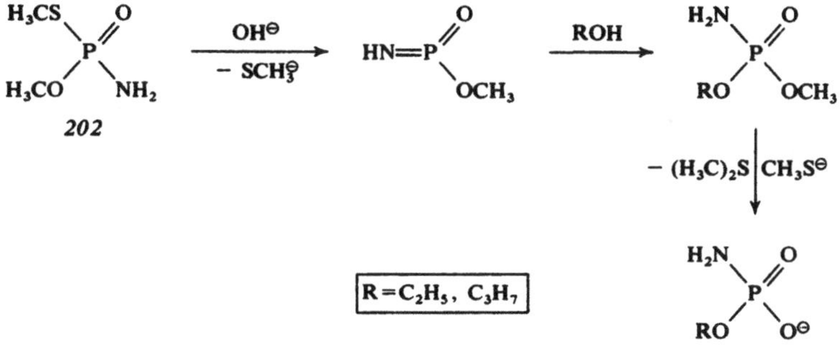

202

Evidence for an intermediate metaphosphorimidate is also provided by a three-phase experiment [110b,139]: treatment of a suspension of the two functionalized polymers *203* and *205* in dioxane with proton sponge leads to formal transfer of a phosphorodiamidate moiety from *203* to *205* with formation of *206*. It appears attractive to assume that the aminoiminooxophosphorane *204* functions as a phosphorylating reagent which arises by elimination from *203*. The question remains open whether it exists in solution as free monomer, as oligomer, or in the form of a dioxane complex.

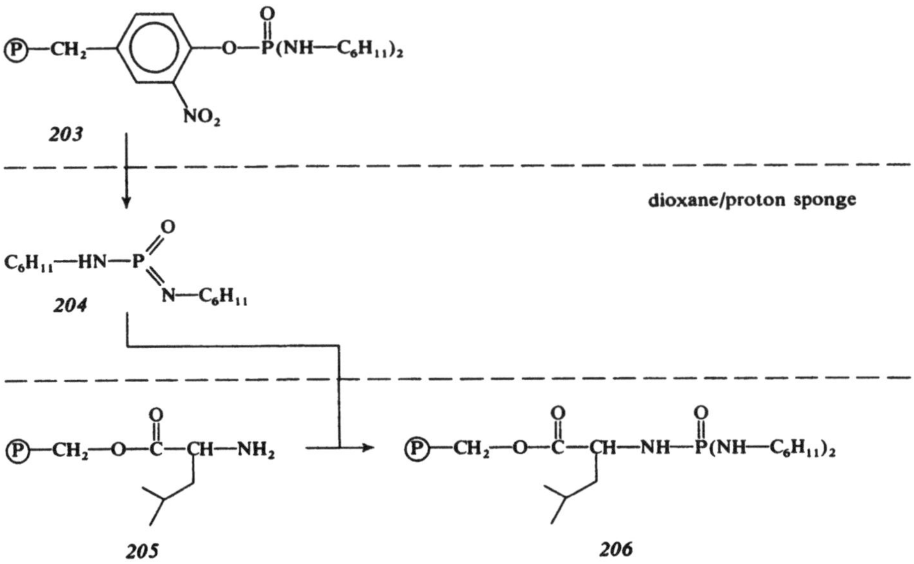

In accord with these mechanistic ideas, photolysis of phosphoryl azides of type $(RO)_2PO—N_3$ in cyclohexane gives predominantly insertion products. Expectedly, no evidence is obtained for the intermediacy of a corresponding metaphosphate since 1,2-OR shifts do not generally occur.

Photolysis of bis(dimethylamino)phosphoryl azide *207* [140] represents an entirely different entry to a metaphosphorimidate. If the reaction is performed in cyclohexane, it gives only 7% of the amide *209* which can be rationalized as the insertion product of the intermediate nitrene *208* into a CH bond of cyclohexane. The major product component is a polymer. The assumption that it is polymeric aminometaphosphorimidate *212* is substantiated indirectly by the nature of the principal product of photolysis of *207* in methanol. A 1,2-shift of a NMe_2 moiety which

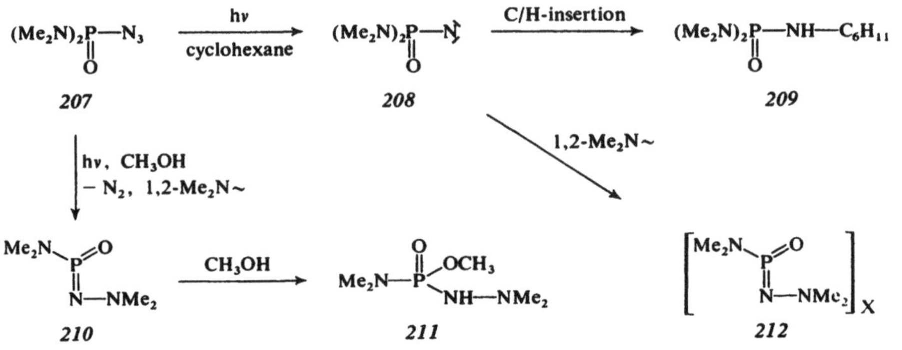

accompanies N_2 elimination leads to the metaphosphorimidate *210* which forms the observed final product *211* by addition of methanol.

Note Added in Proof

A new class of metaphosphinates, namely iminomethylenephosphoranes (*1* with $R = N(SiMe_3)_2$, $X = NSiMe_3$) is accessible by reaction of diazoalkanes with iminophosphines $(SiMe_3)_2N{-}P{=}N{-}SiMe_3$. Depending on the substituents on the methylene carbon atom, compounds of this class are either stable or dimerize across the P=C or P=N double bond [141].

Methyleneoxophosphorane *9*, generated thermally from diazocompound 7, has been trapped by [4+2]-cycloaddition with an α-acylated ketene [142].

The hydrolysis of some phenylphosphonamides, $R_2N{-}P({=}O)(C_6H_5)O^{\ominus}$, in their protonated forms has been investigated. All evidence points to a $S_N2(P)$ mechanism rather than a fragmentation into R_2NH and the monomeric $C_6H_5{-}PO_2$ (*23*) [143].

Phosphoryl azides seem to provide a general access to the highly reactive iminooxophosphoranes, as has been seen for the sequence *78*→*79*→*80*. Thus, irradiation of diphenylphosphoryl azide *91* at 253.7 nm yields monomeric *92* which dimerizes to a phosphadiazetidine in the absence of trapping agents. *92* can also be trapped by alcohols, amines, methyl iodide or epoxides [144]. In a similar manner, photolysis of (di-tert-butyl)phosphoryl azide or (di-isopropyl)phosphoryl azide in methanol leads to the phosphonamidate $R{-}P(O)(OCH_3)NHR$ with $R = n\text{-}C_4H_9$ or $i\text{-}C_3H_7$, respectively. Once again, the intermediacy of a monomeric iminooxophosphorane is assumed [145].

Full details on the phosphorylation of water and alcohols by 4-nitrophenyl dihydrogen phosphate and the $N(C_4H_9)^{\oplus}$- and $N(CH_3)_4^{\oplus}$-salts of its mono- and dianion have been published [146]. Phosphoryl group transfer from the monoanion and dianion is thought to proceed via the monomeric $PO_3^{\ominus}$ ion. Addition of the sterically unhindered amine quinuclidine to an acetonitrile solution containing the phosphate monoanion and tert-butanol produces t-butyl phosphate at a faster rate than does the addition of the more hindered diisopropylethylamine. This nucleophilic catalysis of the phosphorylation reaction is also explained by the intermediacy of the $PO_3^{\ominus}$ ion.

When monomeric metaphosphate anion $PO_3^{\ominus}$ (*102*) is generated form the phosphonate dianion *170* in the presence of the hindered base 2,2,6,6-tetramethylpiperidine, it undergoes reaction with added carbonyl compounds [147]. Thus, it phosphorylates acetophenone to yield the enol phosphate, whereas in the presence of acetophenone and aniline the Schiff base is formed from both compounds, probably by way of the intermediate $C_6H_5{-}C(CH_3)(OPO_3^{2\ominus})(^{\oplus}NH_2C_6H_5)$. This reactivity pattern closely resembles that of monomeric methyl metaphosphate *151* (see Sect. 4.4.2).

When anhydrous Na_3PO_4 is vaporized at 1550–1650 K and the products are condensed in an argon matrix at ca. 10 K, the i.r. spectrum shows that the trapped species is $NaPO_3$ having the bi-dentate C_{2v} structure 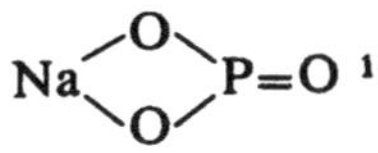 [148].

5 References

1. Summary see Niecke, E., Scherer, O. J.: Nachr. Chem. Techn. *23*, 395 (1975)
2. Niecke, E., Flick, W.: Angew. Chem. *86*, 128 (1974); Angew. Chem., Int. Ed. Engl. *13*, 134 (1974).
3. Scherer, O. J., Kuhn, N.: Chem. Ber. *107*, 2123 (1974).
4. Markovski, L. N., Romanenko, V. D., Ruban, A. V.: Synthesis *1979*, 811
5. Scherer, O. J., Kuhn, N.: J. organomet. Chem. *82* (1974), C3–C6
6. Niecke, E., Wildbredt, D.-A.: Chem. Ber. *113*, 1549 (1980)
7. Wiseman, J., Westheimer, F. H.: J. Am. Chem. Soc. *96*, 4262 (1974); see the references mentioned there, too
8. Pohl, S., Niecke, E., Krebs, B.: Angew. Chem. *87*, 284 (1975); Angew. Chem., Int. Ed. Engl. *14*, 261 (1975)
9. Regitz, M.: Diazolkane, 1. Aufl., S. 78, 115 und 163ff., Thieme, Stuttgart 1977
10. Regitz, M.: Angew. Chem. *87*, 259 (1975); Angew. Chem., Int. Ed. Engl. *14*, 222 (1975)
11. Regitz, M., Liedhegener, A., Anschütz, W., Eckes, H.: Chem. Ber. *104*, 2177 (1971)
12. Jones, M.: personal communication, Princeton University, 1975
13. Regitz, M., Anschütz, W., Bartz, W., Liedhegener, A.: Tetrahedron Lett. *1968*, 3171
14. Regitz, M., Scherer, H., Anschütz, W.: Tetrahedron Lett. *1970*, 753
15. Jugelt, W., Schmidt, D.: Tetrahedron *25*, 5569 (1969)
16. Hufnagel, M.: Ph. D. Thesis, University of Kaiserslautern 1979
17. Regitz, M., Scherer, H., Illger, W. Eckes, H.: Angew. Chem. *85*, 1115 (1973); Angew. Chem., Int. Ed. Engl. *12*, 1010 (1973)
18. Regitz, M., Eckes, H.: Chem. Ber. *113*, 3303 (1980)
19. Eckes, H., Regitz, M.: Tetrahedron Lett. *1975*, 447
20. Regitz, M., Eckes, H.: Tetrahedron *36*, 1039 (1981)
21. Bestmann, H. J., Roth, K., Wilhelm, E., Böhme, R., Burzlaff, H.: Angew. Chem. *91*, 945 (1979); Angew. Chem., Int. Ed. Engl. *18*, 876 (1979)
22. Eckes, H.: Ph. D. Thesis, University of Kaiserslautern 1974
23. Urgast, K.: Ph. D. Thesis, in preparation, University of Kaiserslautern
24. Regitz, M., Bartz, W.: Chem. Ber. *103*, 1477 (1970); s. auch Lit. [9] S. 42ff.
25. Staudinger, H., Endle, R.: Liebigs Ann. Chem. *401*, 263 (1913)
26. Gompper, R., Studeneer, A., Elser, W.: Tetrahedron Lett. *1968*, 1019; Osao, T., Morita, N., Iwagame, N.: Bull. Chem. Soc. Jpn. *47*, 773 (1974)
27. Regitz, M., Anschütz, W.: Chem. Ber. *102*, 2216 (1969)
28. Horner, L., Schwarz, H.: Tetrahedron Lett. *1966*, 3579
29. Horner, L., Bauer, G.: Tetrahedron Lett. *1966*, 3573
30. Martin, M.: Ph. D. Thesis, University of Kaiserslautern 1978
31. Hayes, G., Holt, G.: J. Chem. Soc., Perkin Trans. I *1973*, 1206
32. Regitz, M., Bennyarto, F., Heydt, H.: Liebigs Ann. Chem. *1981*, in press
33. Bennyarto, F.: Diploma Thesis, University of Kaiserslautern 1978
34. Martin, M., Regitz, M., Maas, G.: Liebigs Ann. Chem. *1978*, 789
35. Regitz, M., Illger, W., Maas, G.: Chem. Ber. *111*, 705 (1978)
36. Maas, G., Regitz, M., Illger, W.: Chem. Ber. *111*, 726 (1978)
37. Illger, W.: Ph. D. Thesis, University of Saarbrücken 1975
38. Yoshituji, M., Tagawa, J., Inamoto, N.: Tetrahedron Lett. *1979*, 2415
39. Divisa, B.: Tetrahedron *35*, 181 (1979)
40. Michaelis, A., Rothe, F.: Ber. Dtsch. Chem. Ges. *25*, 1747 (1882)
41. Michaelis, A.: Liebigs Ann. Chem. *293*, 193 (1896); *294*, 1 (1896) sowie *315*, 43 (1901)
42. Cherbuliez, E., Baehler, B., Hunkeler, F., Rabinowitz, J.: Helv. Chim. Acta *44*, 1812 (1961)
43. Connat, J. B., Pollack, S. M.: J. Am. Chem. Soc. *43*, 1665 (1921)
44. Sigal, J., Loew, L.: J. Am. Chem. Soc. *100*, 6394 (1978)
45. Harger, M. J. P.: J. Chem. Soc., Chem. Commun. *1971*, 442
46. Harger, M. J. P.: J. Chem. Soc., Perkin Trans. I, *1974*, 2604
47. Wiseman, J., Westheimer, F. H.: J. Amer. Chem. Soc. *96*, 4262 (1974)
48. Harger, M. J. P., Stephen, M. A.: J. Chem. Soc. *1980*, 705
49. Harger, M. J. P.: J. Chem. Soc., Chem. Commun. *1979*, 930
50. Reichle, W. T.: Inorg. Chem. *3*, 402 (1964)

51. Cates, L. A.: Phosphorus *5*, 1 (1974)
52. Bruice, T. C., Benkovic, S. J.: Bioorganic Mechanisms, Vol. *2*, Chapter 1, W. A. Benjamin, New York 1966
53. Kirby, A. J., Warren, S. G.: The Organic Chemistry of Phosphorus, Chapter 10, Elsevier, Amsterdam 1967
54. Bunton, C. A.: Acc. Chem. Res. *3*, 257 (1970)
55. Benkovic, S. J.: Comprehensive Chemical Kinetics, Vol. *10*, p. 1, Elsevier, New York 1972
56. Benkovic, S. J., Schray, K. J.: The Enzymes, Vol. *8*, p. 201, Academic Press, New York 1973
57. Butcher, W. W., Westheimer, F. H.: J. Amer. Chem. Soc. *77*, 2420 (1955)
58a. Barnard, D. W. C., Bunton, C. A., Llewellyn, D. R., Oldham, K. G., Silver, B. L., Vernon, C. A.: Chem. Ind. *1955*, 760; b. Bunton, C. A., Llewellyn, D. R., Oldham, K. G., Vernon, C. A.: J. Chem. Soc. *1958*, 3574
59. Harvan, D. J., Hass, J. R., Busch, K. L., Bursey, M. M., Ramirez, F., Meyerson, S.: J. Amer. Chem. Soc. *101*, 7409 (1979)
60. Vernon, C. A.: Chem. Soc. Special Publ. *8*, 17 (1957)
61. Bunton, C. A., Mhala, M. M., Oldham, K. G., Vernon, C. A.: J. Chem. Soc. *1960*, 3293
62. Bunton, C. A., Fendler, E. J., Humeres, E., Yang, K.-U.: J. Org. Chem. *32*, 2866 (1967)
63. Bunton, C. A., Llewellyn, D. R., Oldham, K. G., Vernon, C. A.: J. Chem. Soc. *1958*, 3588
64. Kumamoto, J., Westheimer, F. H.: J. Am. Chem. Soc. *77*, 2515 (1955)
65. Kirby, A. J., Varvoglis, A. G.: J. Am. Chem. Soc. *89*, 415 (1967)
66. Gorenstein, D. G.: J. Am. Chem. Soc. *94*, 2523 (1972)
67. Gorenstein, D. H., Lee, Y. G., Kar, D.: J. Am. Chem. Soc. *99*, 2264 (1977)
68. Milstein, S., Fife, T. H.: J. Am. Chem. Soc. *89*, 5820 (1967)
69. Kirby, A. J., Varvoglis, A. G.: J. Am. Chem. Soc. *88*, 1823 (1966)
70. Ramirez, F., Marecek, J. F.: J. Am. Chem. Soc. *101*, 1460 (1979)
71. Ramirez, F., Marecek, J. F.: Tetrahedron *35*, 1581 (1979)
72. Ramirez, F., Marecek, J. F.: Pure Appl. Chem. *52*, 1021 (1980)
73. Kirby, A. J., Varvoglis, A. G.: J. Chem. Soc. *B 1968*, 135
74. Considering however, that dioxane forms a stable complex with SO_3, so a dioxane-$PO_3^{\ominus}$ adduct doesn't seem to be inconceivable
75. Jencks, W. P.: Acc. Chem. Res. *13*, 161 (1980)
76. Disabato, G., Jencks, W. P.: J. Am. Chem. Soc. *83*, 4400 (1961)
77. Park, J. H., Koshland jr., D. E.: J. Biol. Chem. *233*, 986 (1958)
78. Allen jr., C. M., Jones, M. E.: Biochemistry *3*, 1238 (1964)
79. Chanley, J. D., Gindler, E. M., Sobotka, H.: J. Am. Chem. Soc. *74*, 4347 (1952); Chanley, J. D., Gindler, E. M.: J. Am. Chem. Soc. *75*, 4035 (1953); Chanley, J. D., Feageson, E.: J. Am. Chem. Soc. *77*, 4002 (1955)
80. See the discussion in ref. [53], p. 48–54, too
81. Bender, M. L., Lawlor, J. M.: J. Am. Chem. Soc. *85*, 3010 (1963)
82. Benkovic, S. J., Schray, K. J.: Biochemistry *7*, 4090 (1968)
83. Schray, K. J., Benkovic, S. J.: J. Am. Chem. Soc. *93*, 2522 (1971)
84. Benkovic, S. J., Sampson, E. J.: J. Am. Chem. Soc. *93*, 4009 (1971)
85. Hallmann, M., Lapidot, A.: J. Chem. Soc. *1960*, 419
86. Allen, G. W., Haake, P.: J. Am. Chem. Soc. *95*, 8080 (1973)
87. Brown, D. M., Hamer, N. K.: J. Chem. Soc. *1960*, 1155
88. Samuel, D., Silver, B.: J. Chem. Soc. *1961*, 4321
89. Miller, D. L., Ukena, T.: J. Am. Chem. Soc. *91*, 3050 (1969)
90. Miller, D. L., Westheimer, F. H.: J. Am. Chem. Soc. *88*, 1507 (1966)
91. Conant, J. B., Cook, A. A.: J. Am. Chem. Soc. *42*, 830 (1920)
92. Conant, J. B., Coyne, B. B.: J. Am. Chem. Soc. *44*, 2530 (1922)
93. Bermann, E., Bondi, A.: Ber. Dtsch. Chem. Ges. *66*, 278 (1933)
94. Maynard, J. A., Swan, J. M.: Proc. Chem. Soc. (London) *1963*, 61; Aust. J. Chem. *16*, 596 (1963)
95. Kenyon, G. L., Westheimer, F. H.: J. Am. Chem. Soc. *88*, 3561 (1966)
96. a: Loew, L. M.: J. Am. Chem. Soc. *98*, 1630 (1976); b: Loew, L. M., McArthur, W. R.: J. Am. Chem. Soc. *99*, 1019 (1977)
97. Chanley, J. D., Feageson, E.: J. Am. Chem. Soc. *85*, 1181 (1963)
98. Guthrie, J. P.: J. Am. Chem. Soc. *99*, 3991 (1977)

99. Kirby, A. J., Jencks, N. P.: J. Am. Chem. Soc. *87*, 3217 (1965)
100. a: Rebek, J., Gaviña, F.: J. Am. Chem. Soc. *97*, 3221 (1975); b: Rebek, J., Gaviña, F., Navarro, C.: J. Am. Chem. Soc. *100*, 8113 (1978)
101. Kosolapoff, G. H.: Organophosphorus Compounds, p. 347, 352, Wiley, New York 1950
102. Langheld, K.: Ber. Dtsch. Chem. Ges. *43*, 1857 (1910); *44*, 2078 (1911)
103. Steinkopf, W., Schubart, J.: Liebigs Ann. Chem. *424*, 1 (1921)
104. Schramm, G., Wissmann, H.: Chem. Ber. *91*, 1073 (1958)
105. Cramer, F., Hettler, H.: Cnem. Ber. *91*, 1181 (1958)
106. Steinkopf, W.: Liebigs Ann. Chem. *413*, 343 (1917)
107. Kluger, R.: J. Org. Chem. *38*, 2721 (1973)
108. Clapp, C. H., Westheimer, F. H.: J. Am. Chem. Soc. *96*, 6710 (1974)
109. Clapp, C. H., Satterthwait, A. C., Westheimer, F. H.: J. Am. Chem. Soc. *97*, 6873 (1975)
110. Satterthwait, A. C., Westheimer, F. H.: J. Am. Chem. Soc. *100*, 3197 (1978)
111. Meyerson, S., Kuhn, E. S., Ramirez, F., Marecek, J. F., Okazaki, H.: J. Am. Chem. Soc. *102*, 2398 (1980)
112. Meyerson, S., Kuhn, E. S., Ramirez, F., Marecek, J. F., Okazaki, H.: J. Am. Chem. Soc. *100*, 4062 (1978)
113. Tökes, L., Jones G.: Org. Mass Spectrom. *10*, 241 (1975)
114. Stec, W. J., Zielińska, B., Van Wazer, J. R.: Org. Mass Spectrom. *10*, 485 (1975)
115. Maas, G., Hoge, R.: Liebigs Ann. Chem. *1980*, 1028
116. Constenla, M., Dimroth, K.: Chem. Ber. *109*, 3099 (1976)
117. Kirby, A. J., Younas, M.: J. Chem. Soc. B. *1970*, 510
118. Farber, S. J.: J. Org. Chem. *34*, 767 (1969)
119. Satterthwait, A. C., Westheimer, F. H.: J. Am. Chem. Soc. *102*, 4464 (1980)
120. Satterthwait, A. C., Westheimer, F. H.: in Stec, W. J.: Phosphorus Chemistry directed towards Biology, p. 118, Pergamon Press, Oxford/New York 1980
121. Zhdanov, R. J., Zhenodarova, S. M.: Synthesis *1975*, 222
122. Todd, A. R.: Proc. Nat. Acad. Sci. (USA) *45*, 1389 (1957); Proc. Chem. Soc. (London) *1961*, 187; *1962*, 199
123. Zarytova, V. F., Knorre, D. G., Lebedev, A. V., Levina, A. S., Rezvukhin, A. I.: Doklady Akad. Nauk. SSSR *212*, 630 (1973); Chem. Abstr. *79*, 146782j (1973); Knorre, D. G., Lebedev, A. V., Levina, A. S., Rezvukhin, A. I., Zarytova, V. F.: Tetrahedron *30*, 3073 (1974)
124, Knorre, D. G., Zarytova, V. F.: Nucleic Acids Res. *3*, 2709 (1976)
125. Weimann, G., Khorana, H. G.: J. Am. Chem. Soc. *84*, 4329 (1962)
126. Knorre, D. G., Zarytova, V. F., Lebedev, A. V., Khalimskaya, L. M., Sheshegova, E. A.: Nucleic Acids Res. *5*, 1253 (1978)
127. Blackburn, G. M., Brown, M. J., Harris, M. R., Shire, D.: J. Chem. Soc. *C 1969*, 676
128. Jacob, T. M., Khorana, H. G.: J. Am. Chem. Soc. *86*, 1630 (1964)
129. Samuel, D., Westheimer, F. H.: Chem. Ind. (London) *1959*, 51
130. Crunden, E. W., Hudson, R. F.: Chem. Ind. (London) *1958*, 1478; J. Chem. Soc. *1962*, 3591
131. Traylor, P. S., Westheimer, F. H.: J. Am. Chem. Soc. *87*, 553 (1965)
132. Heath, D. R.: J. Chem. Soc. *1956*, 3796, 3804
133. Williams, A., Douglas, K. T.: J. Chem. Soc. Perkin Trans. II *1972*, 1454; *1973*, 318
134. Gerrard, A. F., Hamer, N. K.: J. Chem. Soc. (B) *1968*, 539
135. Westheimer, F. H.: Chem. Soc. Special Publ. *8*, 181 (1957)
136. Gerrard, A. F., Hamer, N. K.: J. Chem. Soc. (B) *1969*, 369
137. Gerrard, A. F., Hamer, N. K.: J. Chem. Soc. (B) *1967*, 1122
138. Fahmy, M. A., Khasawinah, A., Fukuto, T. R.: J. Org. Chem. *37*, 617 (1972)
139. Rebek, J., Gaviña, F.: J. Chem. Soc. *97*, 1591 (1975)
140. Breslow, R., Feiring, A., Herman, F.: J. Am. Chem. Soc. *96*, 5937 (1974)
141. Niecke, E., Wildredt, D.-A.: Chem. Ber. *113*, 1549 (1980)
142. Capuano, L., Tammer, T.: Chem. Ber. *114*, 456 (1981)
143. Rahil, J., Haake, P.: J. Am. Chem. Soc. *103*, 1723 (1981)
144. Bertrand, G., Majoral, J.-P., Baceiredo, A.: Tetrahedron Lett. *1980*, 5015
145. Harger, M. J. P., Stephen, M. A.: J. Chem. Soc. Perkin I *1981*, 736
146. Ramirez, F., Marececk, J. F.: Tetrahedron *36*, 3151 (1980)
147. Satterthwait, A. C., Westheimer, F. H.: J. Am. Chem. Soc. *103*, 1177 (1981)
148. Jenny, S. N., Ogden, J. S.: J. Chem. Soc. Dalton *1979*, 1465

Chemistry of Saturated Bicyclic Peroxides (The Prostaglandin Connection)[1]

Waldemar Adam[1] and Allan John Bloodworth[2]

1 Department of Chemistry, University of Puerto Rico, Rio Piedras, Organic Chemistry Institute, University of Würzburg, Federal Republic of Germany
2 Department of Chemistry, University College London, England

Table of Contents

[1] We dedicate this article to Professor Alwyn G. Davies, University College London

This review is concerned with the synthesis, spectroscopic characterisation, and reactions of saturated bicyclic peroxides. Four years ago such compounds were virtually unknown, yet the first example *2a* had been prepared as early as 1938 by catalytic hydrogenation of the naturally occurring peroxide ascaridole *1* (Eq. 1) [1].

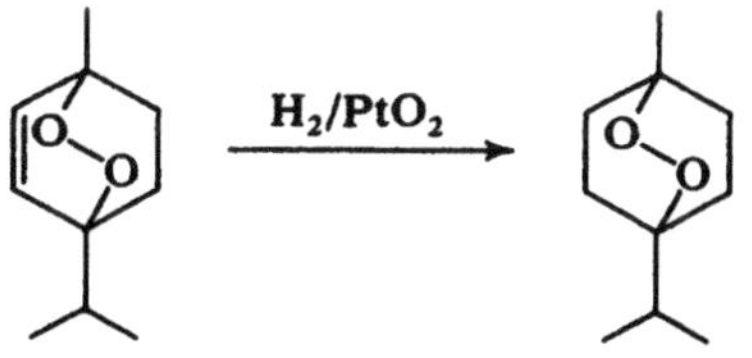

1 $\quad$ *2a* $\qquad$ (1)

The stimulus for the recent surge of activity in this previously dormant area of organic chemistry can be traced to the "prostaglandin connection". That is to the discovery that saturated bicyclic peroxides are key intermediates in the biosynthesis of prostaglandins and other physiologically active substances by the enzymatic oxygenation of polyunsaturated fatty acids.

Although prostaglandins were recognised as constituents of seminal fluid nearly fifty years ago [2], their isolation and structural elucidation were not achieved until the early 1960's [3]. On the basis of structures of the prostaglandins PGE_2 (ether soluble) and $PGF_{2\alpha}$ (phosphate soluble) derived from arachidonic acid *3*, and of molecular oxygen-18 tracer experiments, the prostaglandin endoperoxide *4* was postulated as the logical precursor to these products (Eq. 2) [4].
Confirmation of this suggestion was provided in 1973 by the isolation of *4* from tissue homogenates actively engaged in producing prostaglandins, and the demonstration that it is reduced to $PGF_{2\alpha}$ and isomerised to PGE_2 [5]. More recently thromboxane A_2 [6] and prostacyclin [7], which possess even stronger physiological activity [8] than that of the already potent prostaglandins, have been identified as additional derivatives of the highly labile prostaglandin endoperoxide *4*.

As a reasonable biogenetic pathway for the enzymatic conversion of the polyunsaturated fatty acid *3* into the bicyclic peroxide *4*, the free radical mechanism in Equation 3 was postulated [9]. That such a free radical process is a viable mechanism has been indicated by model studies in which prostaglandin-like products were obtained from the autoxidation of methyl linolenate [10] and from the treatment of unsaturated lipid hydroperoxides with free radical initiators [11].

Similar pathways can be envisaged for the biosynthesis of other prostaglandins and thromboxanes which are derived from polyunsaturated fatty acids closely related to arachidonic acid *3*. Thus substituted 2,3-dioxabicyclo[2.2.1]heptanes such as *4* are key intermediates in the biosynthesis of a large group of hormones, and these substances have wide-ranging biological roles. The central importance of the prostaglandin endoperoxides derives from the fact that they are the point at which the biosynthetic pathway diverges, the plethora of products arising because the peroxidic nucleus can apparently participate in a variety of transformations (Eq. 2). Thus the chemistry associated with the bicyclic peroxide functionality is intimately involved in the biosynthesis. The recognition of this, together with the severe difficulties encountered in attempts to study chemical reactions of the prostaglandin

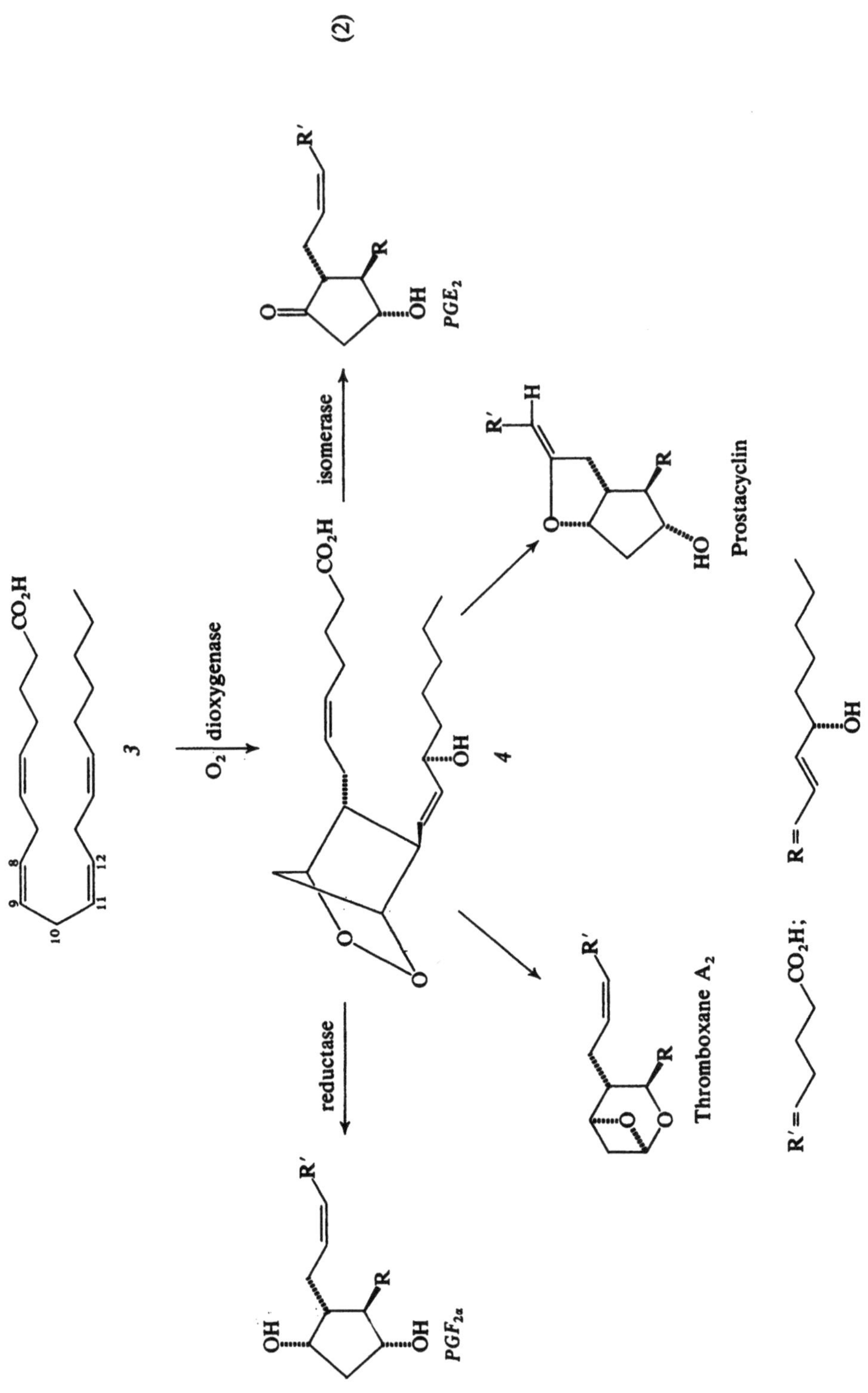

(2)
CO2H
3
O2
dioxygenase
CO2H
OH
4
isomerase
PGE2
reductase
PGF2α
Thromboxane A2
Prostacyclin
R′ = CO2H;
R =
OH

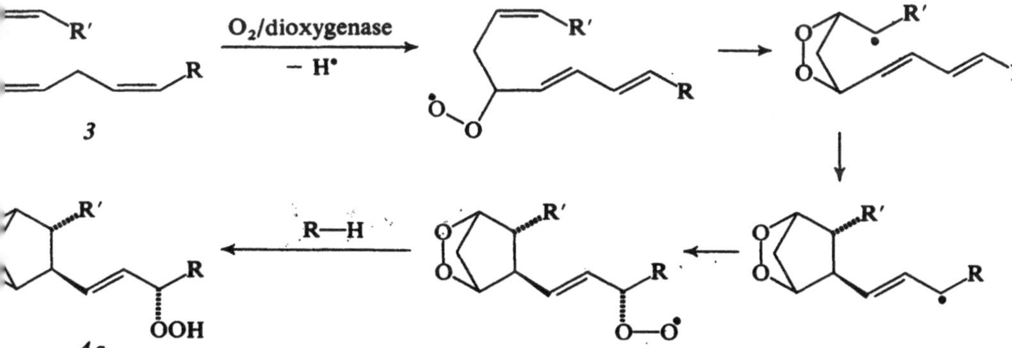

3

4a

(3)

endoperoxides with the minute quantities, instability, and uncertain purity of the samples available from biosynthesis, focused attention on the lack of, and need for, simple chemical models. The chemistry described herein was developed in response to this stimulus.

2,3-Dioxabicyclo[2.2.1]heptane naturally assumed the role of the principal target molecule. It represented a considerable synthetic challenge, for not only is it a strained bicyclic molecule containing the weak and labile O—O bond, but it is also a di(secondary-alkyl) peroxide which is the most difficult type to make by classical procedures [12]. New synthetic methods of exceptional mildness were clearly needed to solve this problem. In the course of the development of such techniques and from a desire to establish their scope, a variety of saturated bicyclic peroxides have been obtained in addition to 2,3-dioxabicyclo[2.2.1]heptane. The question of how substitution patterns and ring sizes affect the reactivity of bicyclic peroxides has further served to broaden interest in the subject.

With synthetic principles now well established, a progress report seems timely. It is hoped that this will encourage further work in the area, particularly on the reactions of bicyclic peroxides where studies are still in their infancy.

1 Synthesis

1.1 Singlet oxygenation with dimide reduction

Endoperoxides *6* are readily available via singlet oxygenation of the corresponding 1,3-cycloalkadienes *5* (Eq. 4) [13].

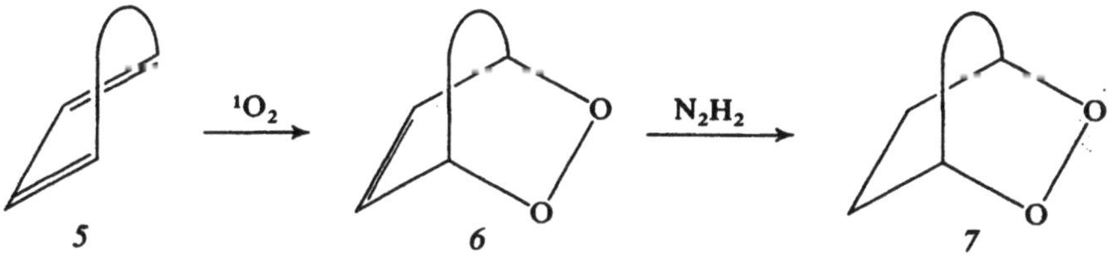

5 *6* *7*

(4)

If the double bond could be reduced without severing the peroxide linkage, such endoperoxides would serve as convenient precursors to the bicyclic peroxides *7*. Unfortunately, the synthesis of dihydroascaridole *2a* (Eq. 1) is exceptional, for catalytic hydrogenation of endoperoxides generally leads to cleavage of the peroxide bond with formation of the saturated *cis*-1,4-diol.

Recently, however, the problem of peroxide cleavage accompanying saturation of the double bond was overcome by engaging the double bond-specific reducing agent diimide [14]. Thus, the endoperoxide of 1,4-diphenylcyclopentadiene *8a* gave a high yield of the bicyclic peroxide *9a* on treatment with diimide, generated by acidification of dipotassium azodicarboxylate (DPAD) with acetic acid in methanol (Eq. 5) [15].

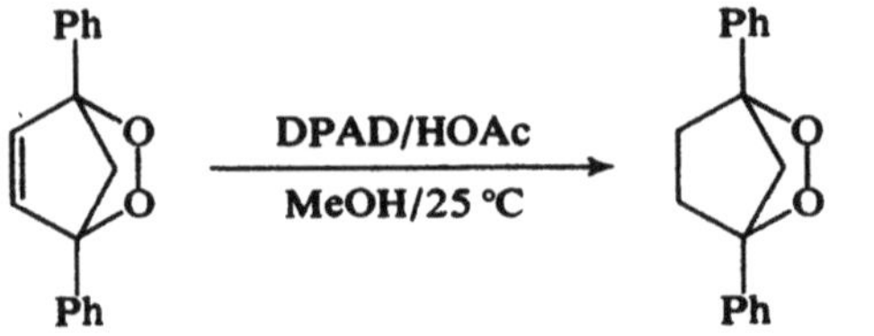

8a *9a* (5)

Subsequently we demonstrated the generality and effectiveness of this synthetic approach to bicyclic peroxides. Among the basic skeletons that have been prepared, figure the [2.2.1]-, [2.2.2]-, [3.2.2]-, [4.2.1]- and [4.2.2]-bicycloperoxides, whose structures are shown below.

We shall now take up each skeleton and its derivatives in detail.

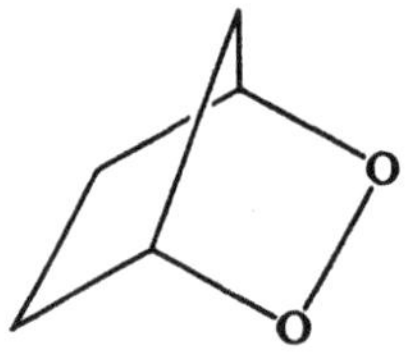

[2.2.1]

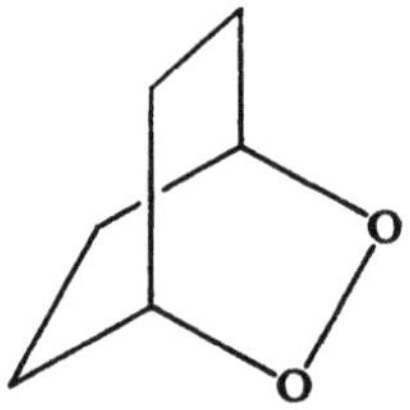

[2.2.2]

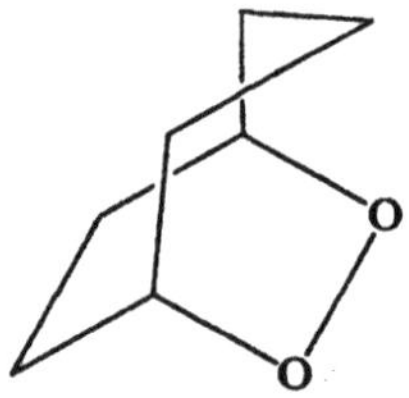

[3.2.2]

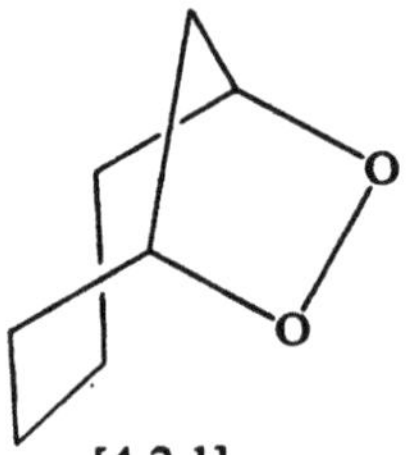

[4.2.1]

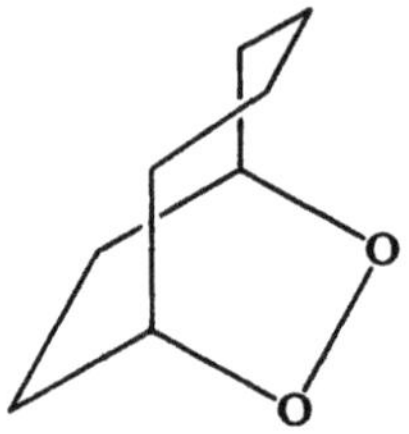

[4.2.2]

1.1.1 The Bicyclo[2.2.1]-skeleton

Among the most difficult bicyclic peroxides to prepare were those possessing the [2.2.1]-skeleton. Although the singlet oxygen-diimide route was first demonstrated for the 1,4-diphenyl derivative *9a*,[15] the parent system *9* could not be prepared because in methanol the endoperoxide *8* rearranged into a complex product mixture even at −30 °C. However, by replacing the polar and protic methanol with the aprotic dichloromethane or fluorotrichloromethane as solvent and by administering stoichiometric amounts of acetic acid at −50 °C, a reasonable yield (ca. 30–40%) of the parent 2,3-dioxabicyclo[2.2.1]heptane *9* could be realised (Eq. 6)[16].

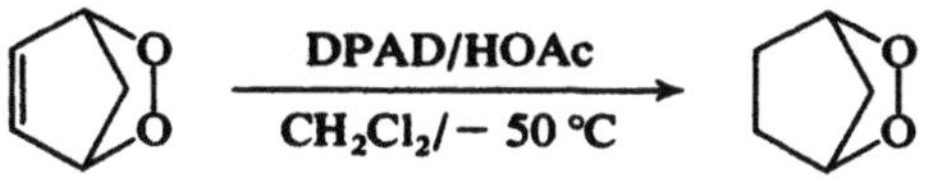

8 *9* (6)

This important modification enabled us to prepare a number of bicyclic peroxides possessing the [2.2.1]-skeleton, of which the important ones are shown below.

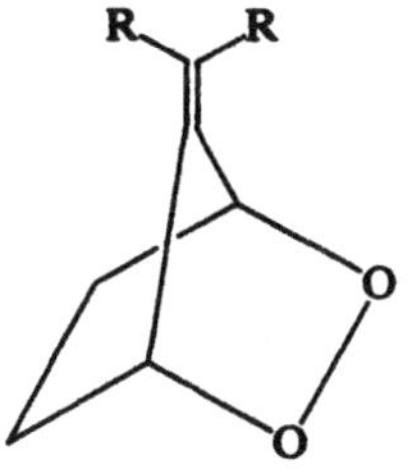

10

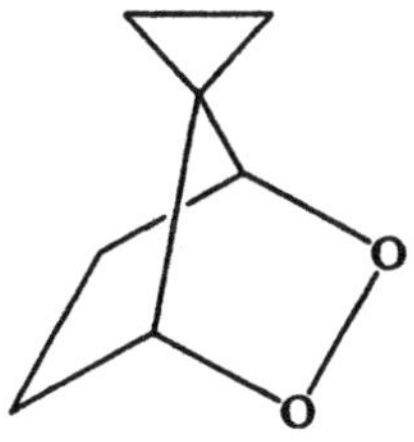

11

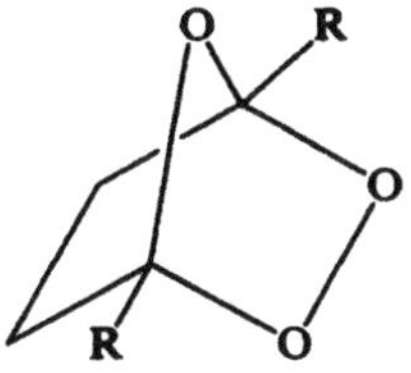

12

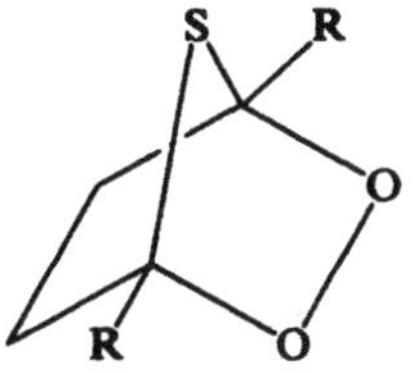

13

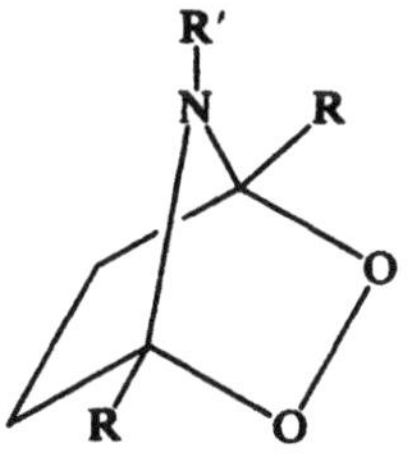

14

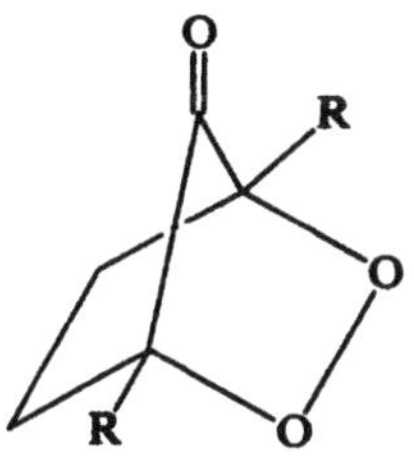

15

Since the corresponding endoperoxide precursors are all too unstable for isolation, the diimide reduction constitutes an important chemical structure confirmation of these elusive intermediates that are obtained in the singlet oxygenation of the respective 1,3-dienes. However, the aza-derivative *14* and the keto-derivative *15* could not be prepared,[17] because the respective endoperoxides of the pyrroles[18] and cyclopentadienones suffered complex transformations even at −50 °C, so that the trapping by the diimide reagent was ineffective.

Of the substrates that have worked well, let us first illustrate the 7-alkylidene-2,3-dioxabicyclo[2.2.1]heptane system *10*. It was known that fulvenes react with singlet oxygen at low temperatures to afford the corresponding endoperoxides; however, attempts to isolate these labile compounds led to decomposition, although NMR identification was possible at —70 °C [19]. When reduction of the singlet oxygenates with diimide was performed at —50 °C, the bicyclic peroxides *10* were obtained in high yield (Eq. 7) [20].

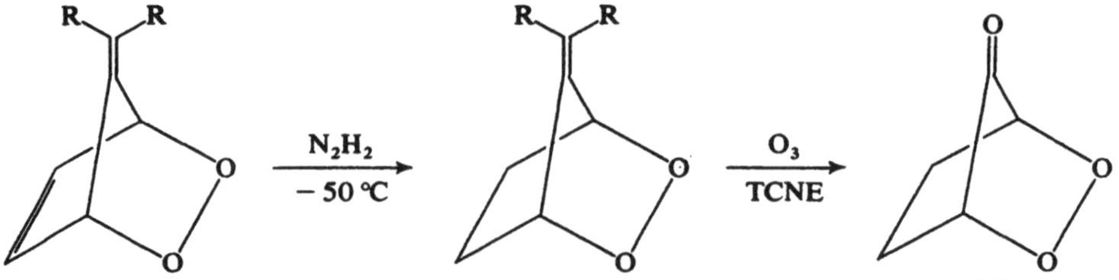

10
(R=Me, Ph)

15

(7)

Furthermore, ozonolysis in the presence of tetracyanoethylene (TCNE) afforded the novel bicyclic peroxide *15* which, as stated already, could not be prepared via the singlet oxygen-diimide route starting from cyclopentadienone. Peroxide *15* was too unstable for isolation, but the characteristic proton resonances at δ 2.0 (m, 4 H) and 4.38 (m, 2 H) ppm are consistent with the assignment.

The bicyclic peroxide *11* was prepared via diimide reduction of the endoperoxide derived from spirocyclopentadiene (Eq. 8) [21]. As before, at elevated temperature the labile endoperoxide rearranges into diepoxide and ketoepoxide, [22] but diimide reduction at —78 °C allows trapping leading to the highly strained bicyclic peroxide *11*.

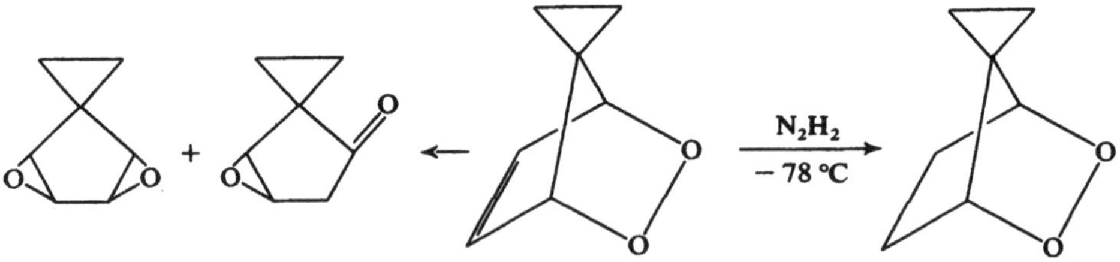

11

(8)

A number of the bicyclic ozonides *12* were prepared in good yield (45–65%) by diimide reduction of furan singlet oxygenates (Eq. 9) [23]. Again, low temperatures were essential because the furan endoperoxides readily transform into 1,2-diacylethylenes. Of course, the bicyclic ozonides *12* can alternatively be prepared via ozonolysis of the appropriate 1,2-disubstituted cyclobutene [24].

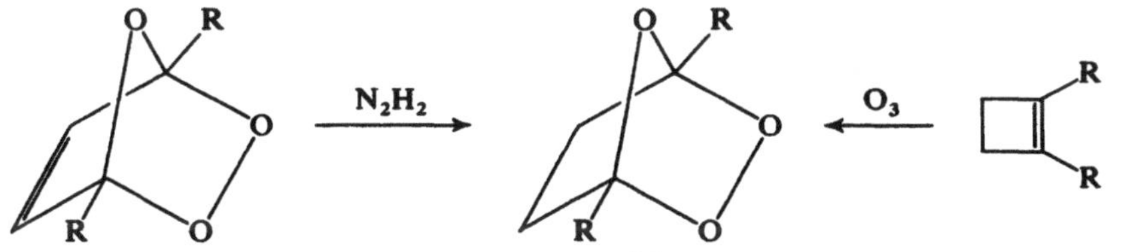

(9)

$R = H, Me, Bu^t, Ph$ 12

On the other hand, the thiaozonides *13* were unknown but could be prepared analogously via singlet oxygenation of 2,5-disubstituted thiophenes and subsequent diimide reduction at low temperature (Eq. 10)[25].

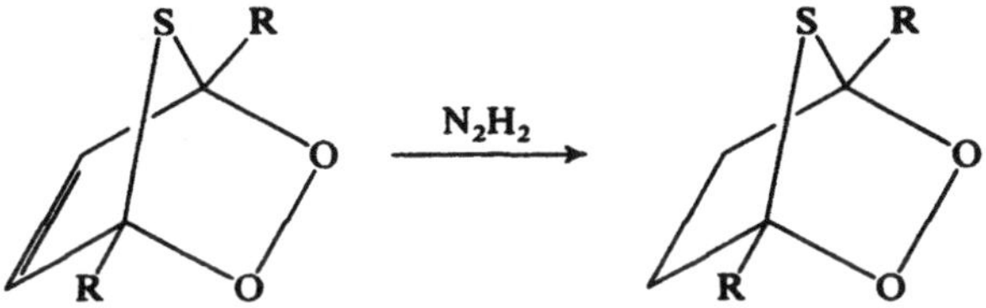

(10)

$R = Me, Bu^t$ 13

The thiophene endoperoxides are considerably more labile than the furan endoperoxides[26], but diimide is sufficiently reactive even at —30 to —50 °C to trap them in the form of thiaozonides *13*.

1.1.2 The Bicyclo[2.2.2]-Skeleton

In addition to the parent compound *2*, the derivatives *2a, b*, the benzo-system *16*, the lactone-peroxides *17*, and the fused polycyclic derivatives *18* and *19* could be prepared via the singlet oxygen-diimide route. For example, the parent system *2* was obtained in *ca.* 40% yield by diimide reduction of the stable 1,3-cyclohexadiene endoperoxide in MeOH at 0 °C[27,28]. Dihydroascaridole *2a* and dihydroergosterol endoperoxide

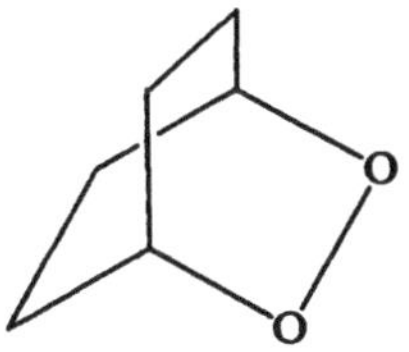

2

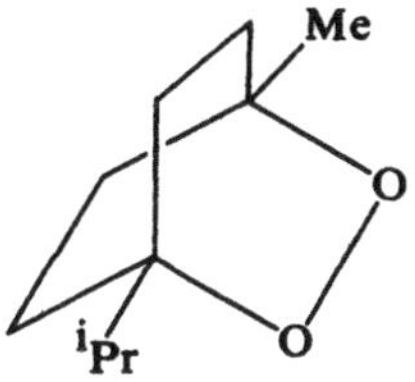

2a

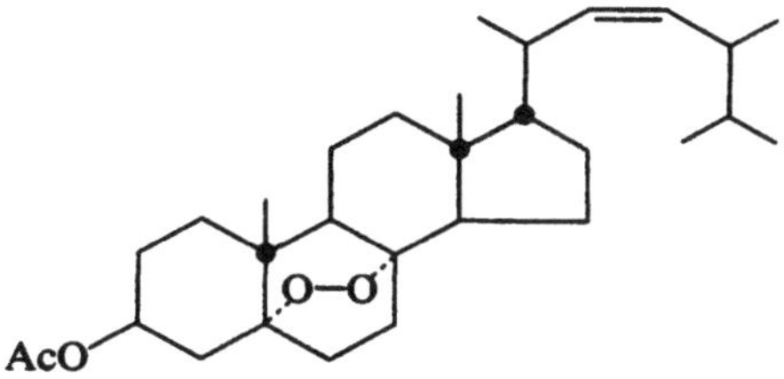

2b

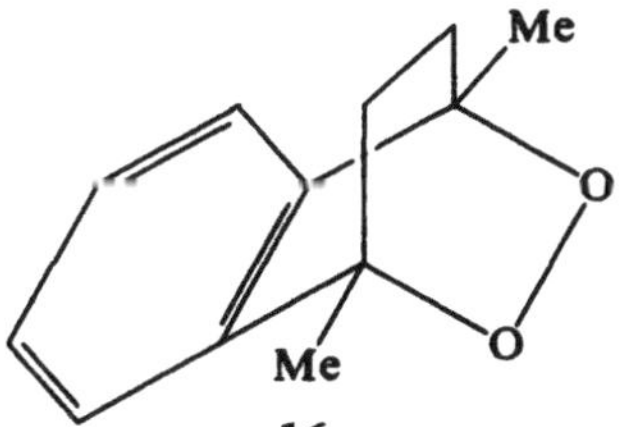

16

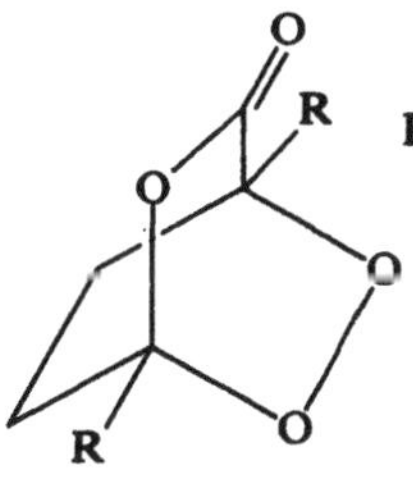

17

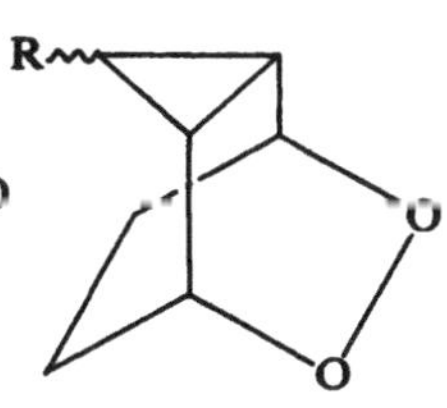

18

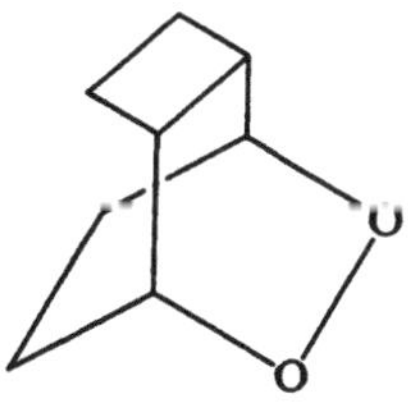

19

2b were prepared similarly [27]. In the latter case, it is clearly evident that the more strained ring-double bond is selectively reduced by diimide, leaving the side chain-double bond intact.

The benzo-derivative *16* is accessible through 1,4-dimethylnaphthalene, which on singlet oxygenation leads to the thermally labile naphthalene-1,4-endoperoxide. This endoperoxide expels singlet oxygen at *ca.* 10 °C, but diimide reduction below 0 °C in MeOH affords the stable dihydro derivative *16* (Eq. 11).

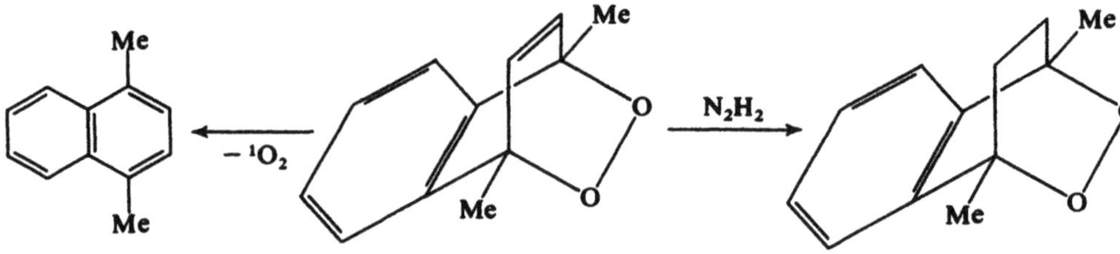

16 (11)

The lactone-peroxides *17* are derived from the corresponding α-pyrones. Singlet oxygenation at low temperature affords the unstable α-pyrone endoperoxides which, on warming up, readily decarboxylate into 1,2-diacylethylenes. However, subambient diimide reduction leads to the desired lactone peroxides *17* (Eq. 12) [29].

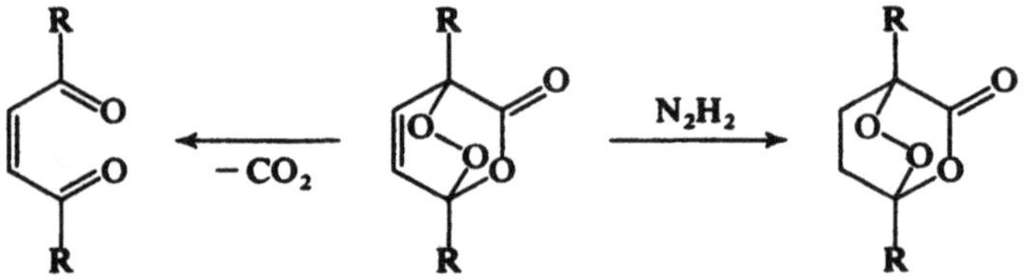

R=H, Ph *17* (12)

The norcarane peroxides *18* are derived from 7-substituted 1,3,5-cycloheptatriene via the singlet oxygen-diimide route (equation 13) [30].

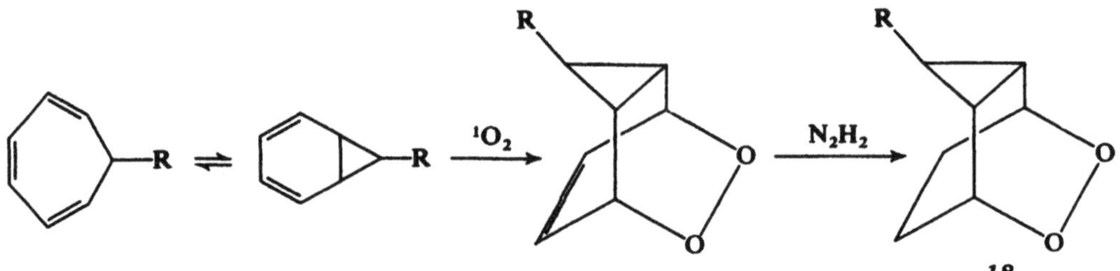

18

R=H, Me, Et, ^{i}Pr, ^{t}Bu, Ph, CN, CO_2Me, CHO

(13)

Depending on the 7-substituent, the norcaradiene endoperoxide can be the major cycloadduct with singlet oxygen. For example, whereas the parent system (R = H) affords only about 4% of the norcaradiene endoperoxide [30a], the 7-cyano derivative leads quantitatively to the norcaradiene adduct [30b)]. Diimide reduction proceeds in most cases quantitatively at 0 °C even in methanol, thus providing a convenient entry into this series of complex bicyclic peroxides.

Similarly, the cyclobutane-fused bicyclic peroxide *19* was prepared by diimide reduction of the corresponding bicyclic endoperoxide derived from 1,3,5-cyclooctatriene (Eq. 14) [31a)].

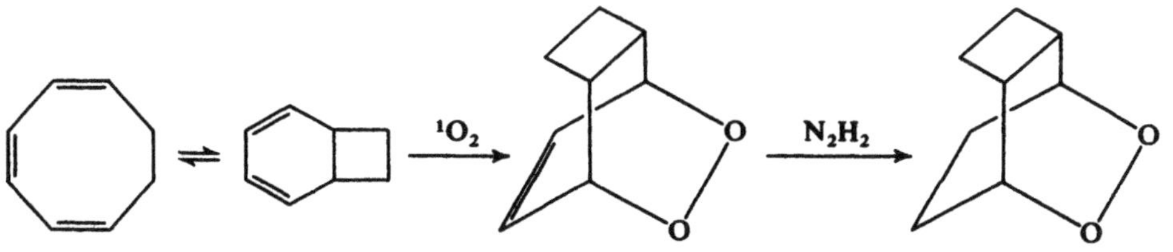

19

(14)

Again, the bicyclic valence isomer coexists in sufficient concentration, that the bicyclic peroxide *19* was readily accessible in ca. 20% yield. Alternatively, the thermally labile bicyclic valence isomer of cyclooctatetraene, namely bicyclo[4.2.0]-octa-2,4,7-triene, was converted into the corresponding endoperoxide on low temperature singlet oxygenation and reduced with diimide to yield *19*.

1.1.3 The Bicyclo[3.2.2]-Skeleton

The parent system *20* was prepared from the 1,3-cycloheptadiene endoperoxide (Eq. 15) in 38% yield [28, 32)]. However, this double bond is quite sluggish towards saturation with diimide, so that a large excess of the diimide reagent is necessary, preferably recycling the incompletely reduced reaction mixture several times.

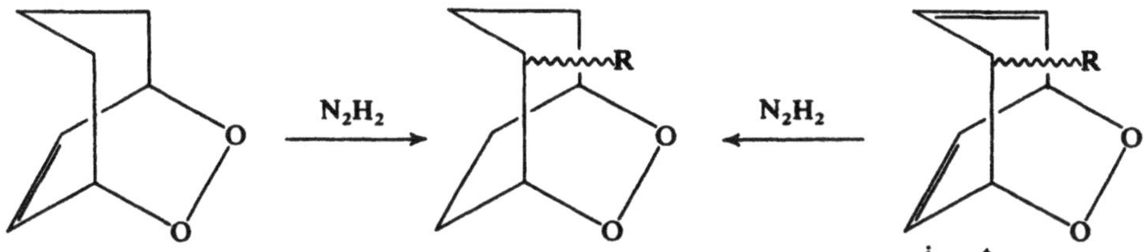

20 (R=H, Me, Et, ^{i}Pr, ^{t}Bu, Ph, OMe)

(15)

Alternatively, the (2 + 4)-tropilidene endoperoxide, which is the major product in the singlet oxygenation of cycloheptatriene [30a)] affords on diimide reduction the desired bicyclic peroxide *20*. The double bond in the two-carbon bridge is reduced selectively, but on exhaustive treatment with excess diimide, the fully saturated substance is obtained. A number of substituted derivatives have been prepared in this way [30)].

The keto-derivative *21* is of interest because the relatively unreactive 3,5-cycloheptadienone substrate, which towards most dienophiles reacts with double bond isomerization, affords the desired endoperoxide (Eq. 16)[33]. Diimide reduction proceeds smoothly, leading to the keto-peroxide *21* in over 90% yield.

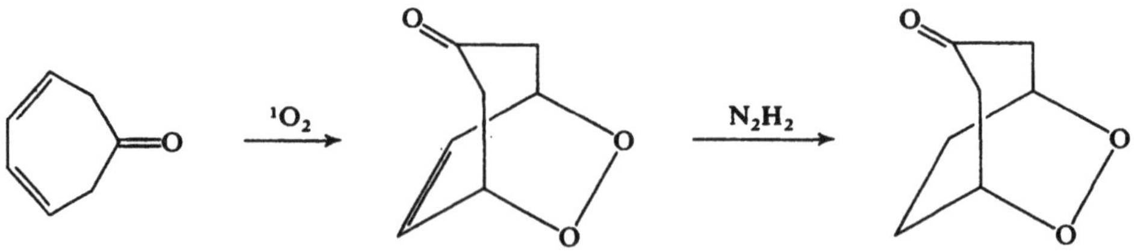

21

(16)

1.1.4 The Bicyclo[4.2.1]-Skeleton

Only one example is known so far, namely the parent compound *22*. As already mentioned, cycloheptatriene affords the norcaradiene endoperoxide (Eq. 13) and the tropilidene (2 + 4)-endoperoxide (Eq. 15); however, another major product is the tropilidene (2 + 6)-endoperoxide (Eq. 17)[30a,d,e].

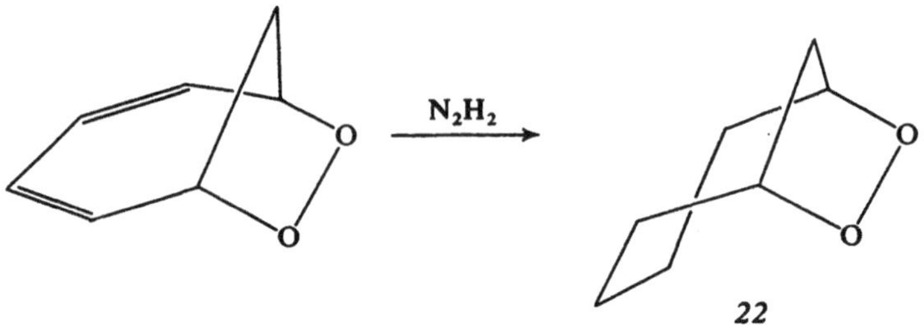

(17)

22

The strained dienic endoperoxide is readily reduced by diimide, leading to the relatively stable bicyclic peroxide in high yield. Again, aprotic solvents such as CH_2Cl_2 or $CFCl_3$ are essential for the diimide reduction, because in MeOH complex rearrangements take place[30d,e].

1.1.5 The Bicyclo[4.2.2]-Skeleton

Like the [3.2.2]-peroxide, the parent system *23* could be prepared in two ways, either from 1,3-cyclooctadiene[28,34] or from 1,3,5-cyclooctatriene[31a] (Eq. 18).

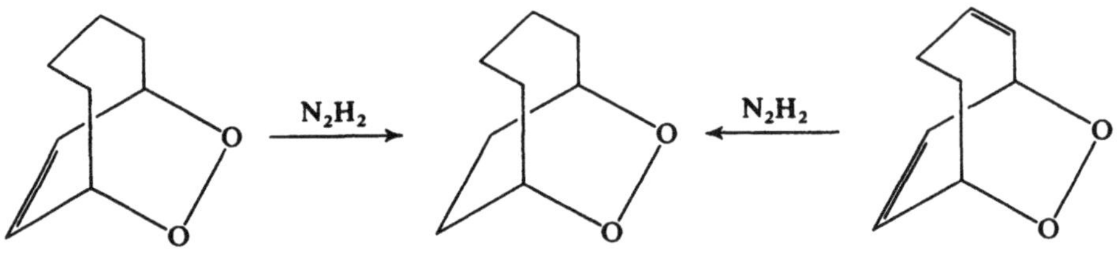

23

(18)

The diimide reduction again proceeds sluggishly and several recycles are essential to achieve complete conversion. The doubly unsaturated endoperoxide is the major product in the singlet oxygenation of 1,3,5-cyclooctatriene (Eq. 14).

1.2 Silver-salt-assisted substitutions

Dialkyl peroxides can be obtained by the alkylation of alkyl hydroperoxides with alkyl halides. Mild conditions can be achieved by assisting the departure of the halide ion with the aid of a suitable silver salt (Eq. 19) [35].

$$\text{HOOR} + \text{R'Hal} + \text{AgX} \rightarrow \text{R'OOR} + \text{AgHal} + \text{HX} \qquad (19)$$

An intramolecular variation of this method, employing 3-bromocycloalkyl hydroperoxides *24*, has proved a versatile procedure for the synthesis of bicyclic peroxides with an [n.2.1] skeletal arrangement, *25* (Eq. 20).

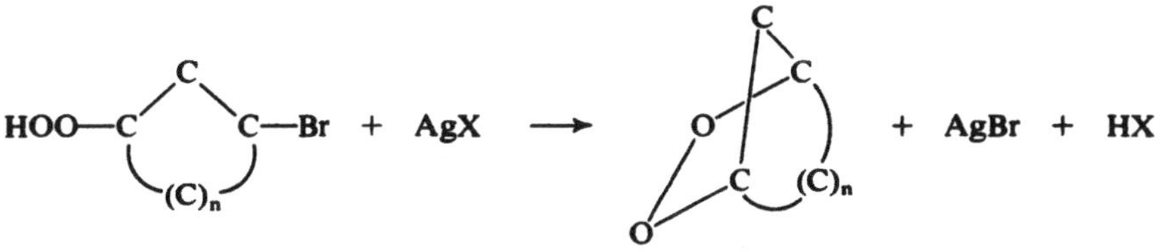

(20)

24 *25*

Derivatives of the [2.2.1]-, [3.2.1]-, [4.2.1]-, and [5.2.1]-bicycloperoxides have been obtained in this way. In describing the results in detail, the expedient of subdividing the discussion according to skeletal type is less appropriate here than an approach based on the methods used to obtain the 3-bromocycloalkyl hydroperoxides *24* needed for the dioxabicyclization.

1.2.1 Using 3-Bromocycloalkyl Hydroperoxides obtained by Peroxybromination of Bicyclopentane

The silver-salt method was used by Porter and Gilmore in one of the first syntheses of 2,3-dioxabicyclo[2.2.1]heptane *9* [36]. Reaction of bicyclopentane with 98% H_2O_2 and *N*-bromosuccinimide (NBS) in ether at —41 °C afforded mainly a 1:1 mixture

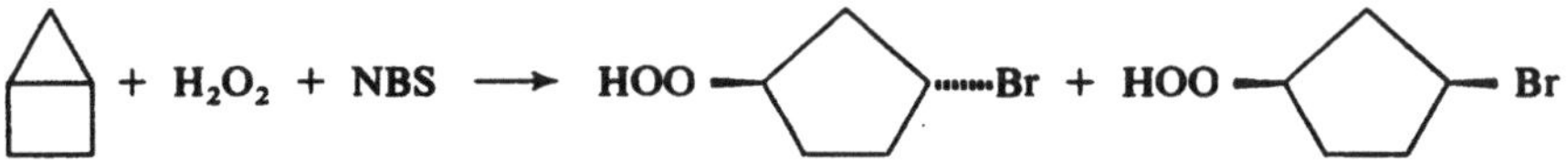

(21)

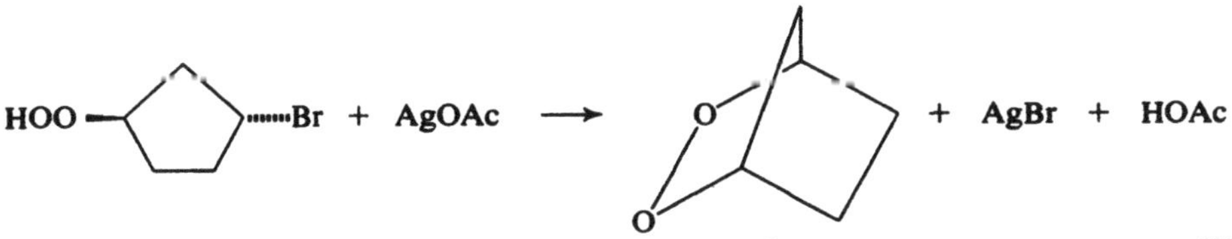

9

(22)

of *cis*- and *trans*-3-bromocyclopentyl hydroperoxide (Eq. 21) which was separated by silica chromatography at —10 °C. Stirring the *trans*-isomer with silver acetate for 30 minutes then gave a quantitative (n.m.r.) yield of *9* (Eq. 22). Under similar conditions, reaction with *cis*-3-bromocyclopentyl hydroperoxide was much slower and not clean, reflecting the known [37] preference for an S_N2 type of transition state for bromide displacement.

1.2.2 Using 3-Bromocycloalkyl Hydroperoxides obtained by Bromination of 2-Cycloalkenyl Hydroperoxides

We incorporated the silver-salt method into a sequence of reactions (Eq. 23) that provides a general route from the commercially available C_5–C_8 cycloalkenes to a series of [n.2.1]-peroxides [38].

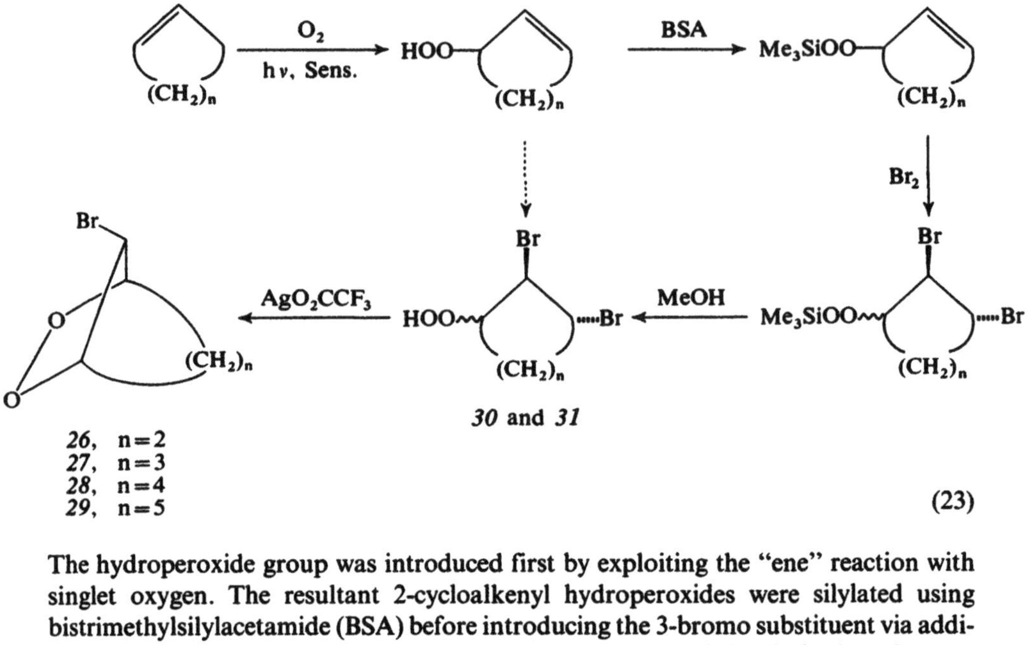

30 and *31*

26, n=2
27, n=3
28, n=4
29, n=5

(23)

The hydroperoxide group was introduced first by exploiting the "ene" reaction with singlet oxygen. The resultant 2-cycloalkenyl hydroperoxides were silylated using bistrimethylsilylacetamide (BSA) before introducing the 3-bromo substituent via addition of bromine. Methanolysis then afforded a mixture of the desired *cis*-2-*trans*-3-dibromocycloalkyl hydroperoxide *30* and the unwanted *trans*-2-*cis*-3-dibromocycloalkyl hydroperoxide *31*. Although bromine can be added directly to the 2-cycloalkenyl hydroperoxides, use of the silylated forms leads to cleaner reactions and a higher proportion of the desired *trans*-3-bromide.

Peroxide ring closures were effected by stirring the 2,3-dibromocycloalkyl hydroperoxides with silver trifluoroacetate, and the bromo-substituted bicyclic peroxides were isolated by silica chromatography at —25 °C. Yields (based on 2-cycloalkenyl hydroperoxide) of 56 and 38% were achieved respectively for the [3.2.1]- and [4.2.1]-compounds, but only 16% of the [2.2.1]- and 13% of the [5.2.1]-peroxide was obtained. The main reason for the low yield of the [2.2.1]-peroxide was that substitution by trifluoroacetate, which competes with the desired dioxabicyclization, is particularly prevalent with the 5-membered ring.

Accordingly, use of silver oxide afforded a vastly improved yield (43%) of *26*. The yield of *29* was similarly improved (to 18%) but here the effect of changing to silver oxide was less dramatic because the main influence is now the fact that, in contrast to the C_5–C_7 dibromocycloalkyl hydroperoxides, the unwanted *31* is the predominant diastereoisomer for the 8-membered ring.

Only one bicyclic peroxide was obtained for each ring system and for *26* and *27* we confirmed that the bromide is *cis* to the peroxide linkage. In the [2.2.1]-compound this was established from its ^{1}H n.m.r. spectrum by the absence of long range W-plan coupling for the C*H*Br proton. The *cis*-configuration of the [3.2.1] compound was indicated by the ^{1}H n.m.r. spectrum of the 2-bromo-3-butoxy-cyclohexanol obtained quantitatively upon reaction with butyllithium (Eq. 24), and has now been confirmed by an X-ray crystal structure determination [39].

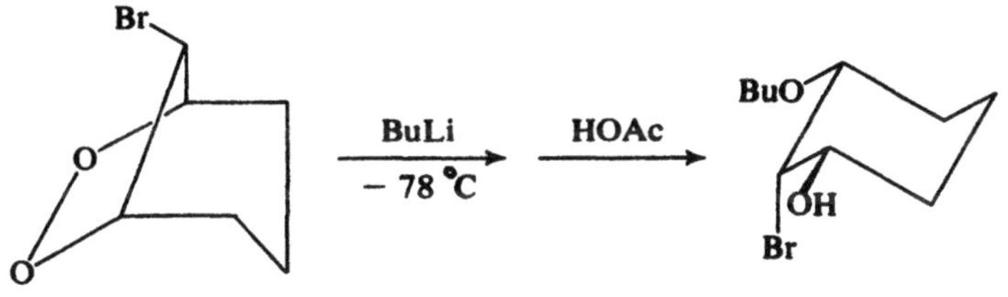

27

(24)

Thus, as expected from earlier work [36,37], the dioxabicyclization proceeds with inversion of configuration at the 3-position of the *cis*-2-*trans*-3-dibromide *30*, the stereochemistry at the 2-position being unaffected. However, experiments with individual diastereoisomers unexpectedly showed that the *trans*-2-*cis*-3-dibromide *31*, also reacts with silver trifluoroacetate, albeit less efficiently, to give the *same* bicyclic peroxide. We feel that this probably proceeds via an isomerisation (Eq. 25).

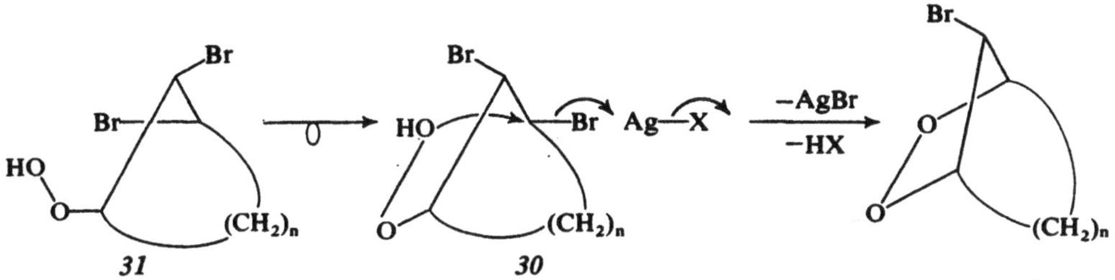

31   30

(25)

1.2.3 Using 3-Bromocycloalkyl Hydroperoxides obtained by Bromination of 3-Cycloalkenyl Hydroperoxides

Interesting results were obtained when 3,4-dibromocycloalkyl hydroperoxides were treated with silver salts.

It occurred to us that bromination of 3-cyclopentenyl hydroperoxide would yield an adduct in which there is always a *trans*-3-bromine, and that ring closure would afford a [2.2.1]-peroxide with a substituent at the 5-position. This additional prostanoid feature would distinguish the compound from existing models (*9a*, *10*, *11*, *15*, and *26*), which, unlike the prostaglandin endoperoxides, were substituted at the bridgehead or 7-position of the [2.2.1]skeleton.

We developed a convenient synthesis of 3-cyclopentenyl hydroperoxide via hydroboration and autoxidation of cyclopentadiene, and bromination proceeded smoothly to afford *32*[40]. Ring closure with silver trifluoroacetate (Eq. 26) afforded a 5-bromo-2,3-dioxabicyclo[2.2.1]heptane *34* (6%) and a 5-trifluoroacetoxy-2,3-dioxabicyclo-[2.2.1]heptane *35* (14%), and it was shown independently that *34* is rapidly converted into *35* by reaction with AgO_2CCF_3. To avoid the trifluoroacetate bromide substitution that accompanies and competes with the dioxabicyclization, *32* was treated with silver oxide and this slowly yielded an isomeric 5-bromo-peroxide *33* (42%) (Eq. 26).

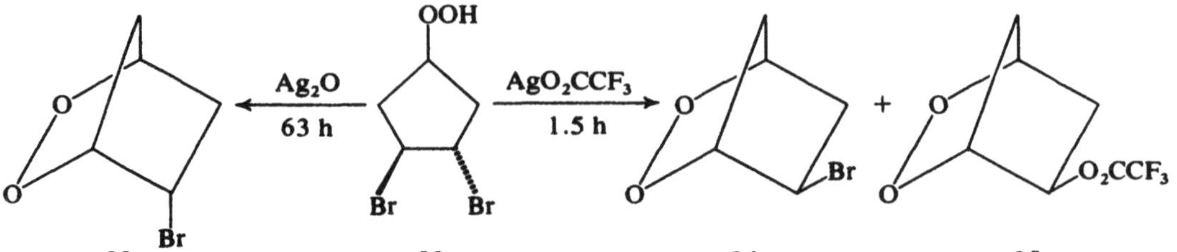

33 *32* *34* *35*

(26)

On the basis of the known preference for displacement of a *trans*-3-bromine with inversion (above), it was originally assumed that *34* had an *endo*-configuration and hence the *exo*-configuration was assigned to *33*, the suggestion being made that *33* was the product of equilibrium control. The correct assignments were established by an independent synthesis of the *exo*-isomer *34* via *trans*-hydroperoxybromination of 3-cyclopentenyl bromide and ring closure with silver oxide (Eq. 27).

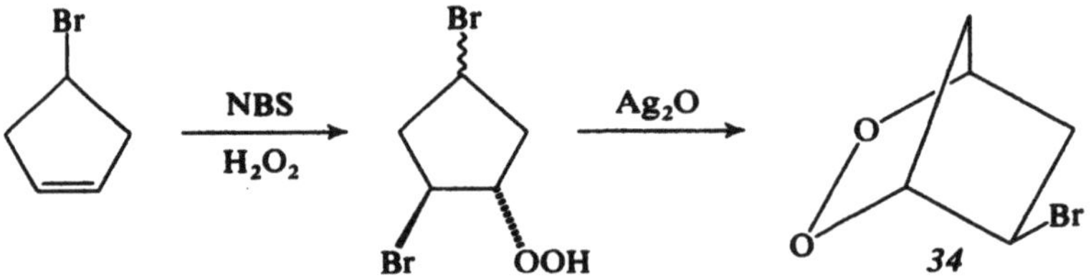

34

(27)

Incidentally, this represents a better route to *34* (11%) than that shown in equation 26, since it is not complicated by concurrent formation of *35*. The configuration of *34* was confirmed by catalytic hydrogenation, for this afforded a 4-bromo-cyclopentane-1,3-diol that was identical with one of the products of *trans*-hydroxy-bromination of 3-cyclopentenyl alcohol (Eq. 28).

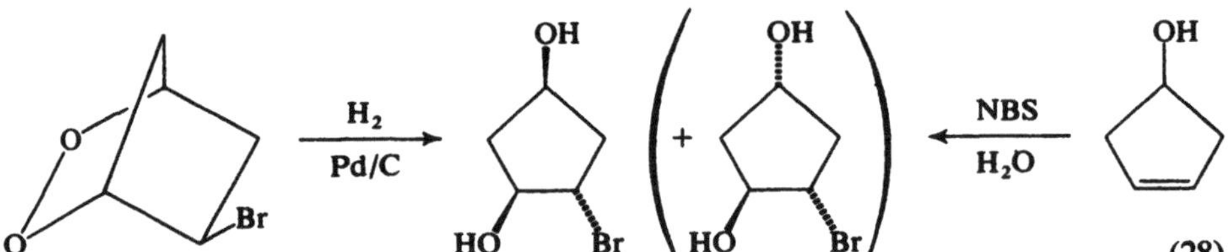

(28)

The possibility that *34* and *35* were formed via *33* was eliminated and hence it must be concluded that, in contradistinction to the reaction with 2,3-dibromocyclopentyl hydroperoxide [38], the AgO_2CCF_3-induced dioxabicyclization of 3,4-dibromocyclopentyl hydroperoxide involves preferential displacement of the *cis*-3-bromine. It seems highly probable that this process is assisted by the *vicinal* bromine, i.e. that the *trans*-bromonium ion *36* is an intermediate. Failure to observe the analogous mechanism with 2,3-dibromocyclopentyl hydroperoxide presumably reflects the disfavoured nature of the mode of ring closure needed in the corresponding species *37*.

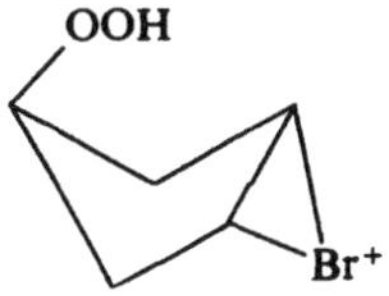

36

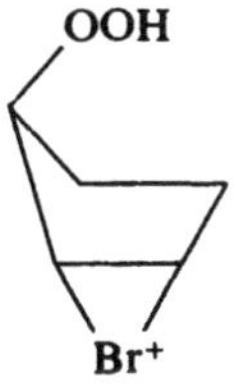

37

To seek further evidence for a bromonium ion-mediated dioxabicyclization and to investigate the regioselectivity of ring closure, we studied reactions with 3,4-dibromocyclohexyl hydroperoxides [41]. We developed a synthesis of 3-cyclohexenyl hydroperoxide based on oxidation of the corresponding N-tosylhydrazine by the procedure of Caglioti et al. [42]. Anisole was the starting material and the full reaction sequence is shown in Eq. 29.

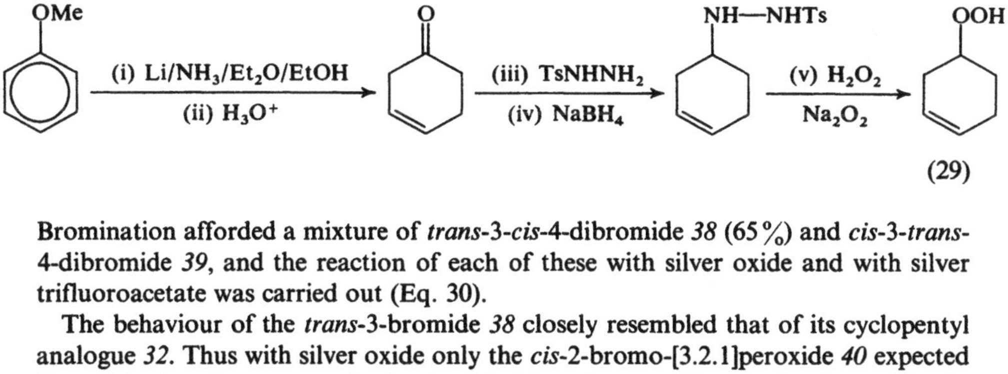

(29)

Bromination afforded a mixture of *trans*-3-*cis*-4-dibromide *38* (65%) and *cis*-3-*trans*-4-dibromide *39*, and the reaction of each of these with silver oxide and with silver trifluoroacetate was carried out (Eq. 30).

The behaviour of the *trans*-3-bromide *38* closely resembled that of its cyclopentyl analogue *32*. Thus with silver oxide only the *cis*-2-bromo-[3.2.1]peroxide *40* expected for a S_N2 ring closure was obtained, and although some *40* was also formed in the reaction of *38* with silver trifluoroacetate, the predominant (90%) bicyclic peroxide obtained was *41*, i.e. the [3.2.1]peroxide available via a bromonium ion mechanism. The behaviour of the *trans*-4-bromide *39* was very revealing, for it did not react with silver oxide and *41* was the only bicyclic peroxide formed with silver trifluoroacetate.

These results support the existence of a bromonium ion pathway for dioxabicyclization of 3,4-dibromocycloalkyl hydroperoxides and confirm a dependence of mechanism upon choice of silver salt. More significantly, from a synthetic viewpoint the results also indicate that both S_N2 and bromonium ion mechanisms are

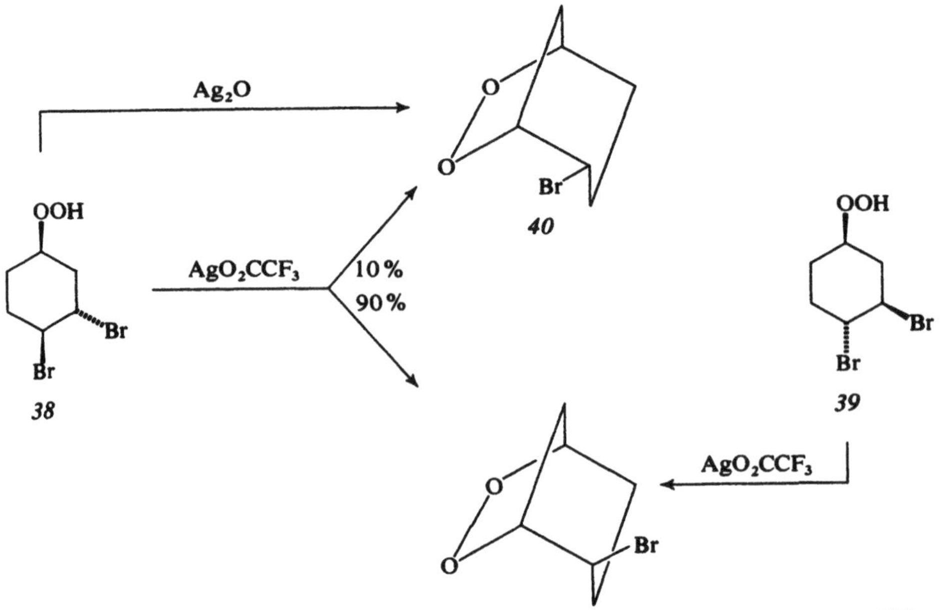

38 40 39 41

(30)

regiospecific. The formation of [3.2.1]-peroxides to the exclusion of [2.2.2]-compounds emphasises the complementary nature of the silver-salt and singlet oxygen-diimide routes to bicyclic peroxides.

1.2.4 Using 3-Bromocycloalkyl Hydroperoxides obtained by Perhydrolysis of 1,3-Dibromocyclopentanes

Since hydrogen peroxide, like alkyl hydroperoxides, can be alkylated by alkyl bromide plus silver trifluoroacetate (Eq. 19, R = H), [35] an attractive variation of the silver-salt-induced dioxabicyclization uses *cis*-1,3-dibromocycloalkane *43* as starting material. Thus Porter and Gilmore obtained 2,3-dioxabicyclo[2.2.1]heptane *9* in 30–40% yield from *cis*-1,3-dibromocyclopentane, which was itself obtained from the corresponding *cis*-diol by reaction with triphenylphosphine dibromide (Eq. 31; R = R′ = H) [36].

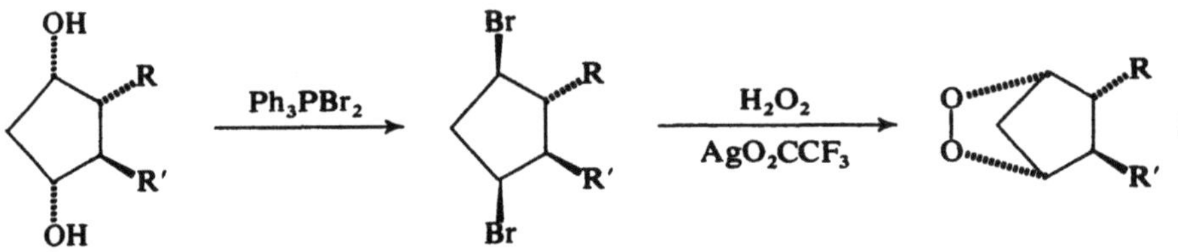

(31)

42 43

This clearly provided a model for the conversion of the prostaglandin $PGF_{2\alpha}$[*42*, R = $CH_2CH\overset{Z}{=}CH(CH_2)_3CO_2H$, R′ = $CH\overset{E}{=}CHCH(OH)C_5H_{11}$] into the corresponding prostaglandin endoperoxide PGH_2 *4*, although an alternative route to dibromide would be required to avoid complications from the presence of the additional OH group at C-15.

The pioneering work in this area was carried out by Johnson et al. [43], who devised a sequence of reactions that converted $PGF_{2\alpha}$ into the key dibromide, 9β, 11β-dibromo-9,11-dideoxy-$PGF_{2\alpha}$ methyl ester, *43a*. Earlier, the same research group had developed a synthesis of dialkyl peroxides based on the reaction of alkyl halides with crown ether — complexed potassium superoxide [44]. By applying this technique to *43a* they achieved the first synthesis of PGH_2 methyl ester, although only about 400 μg were obtained. Porter's group showed that the yield in the dioxabicyclization step could be increased from 3 to 20–25% by replacing superoxide with hydrogen peroxide and silver trifluoroacetate (Eq. 32) [45].

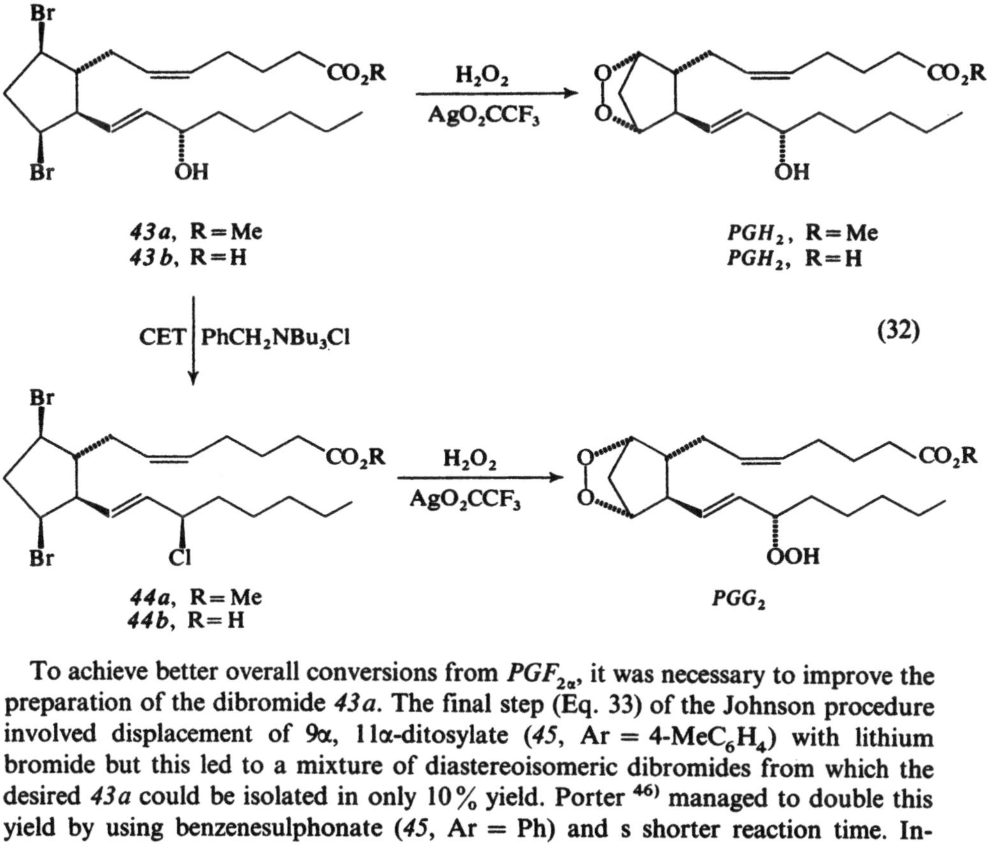

43a, R=Me
43b, R=H

PGH_2, R=Me
PGH_2, R=H

(32)

44a, R=Me
44b, R=H

PGG_2

To achieve better overall conversions from $PGF_{2\alpha}$, it was necessary to improve the preparation of the dibromide *43a*. The final step (Eq. 33) of the Johnson procedure involved displacement of 9α, 11α-ditosylate (*45*, Ar = 4-MeC_6H_4) with lithium bromide but this led to a mixture of diastereoisomeric dibromides from which the desired *43a* could be isolated in only 10% yield. Porter [46] managed to double this yield by using benzenesulphonate (*45*, Ar = Ph) and s shorter reaction time. Incorporating this improvement, the conversion of $PGF_{2\alpha}$ into PGH_2 was achieved for the first time in an overall yield of 2.3% [46]. Hydrolysis of *43a* was carried out using hog pancreas lipase and the best conditions for dioxabicyclization (Eq. 32) of the resultant free acid *43b* were to use a large excess of silver trifluoroacetate and hydrogen peroxide with a reaction time of less than 30 minutes. Porter subsequently reported [47] a much-improved synthesis of *43a* in which the 15-t-butyldimethylsilyl ether *46* [43] was treated with 2-chloro-3-ethylbenzoxazolium tetrafluoroborate (CET) and tetraethylammonium bromide (Eq. 33). This afforded the

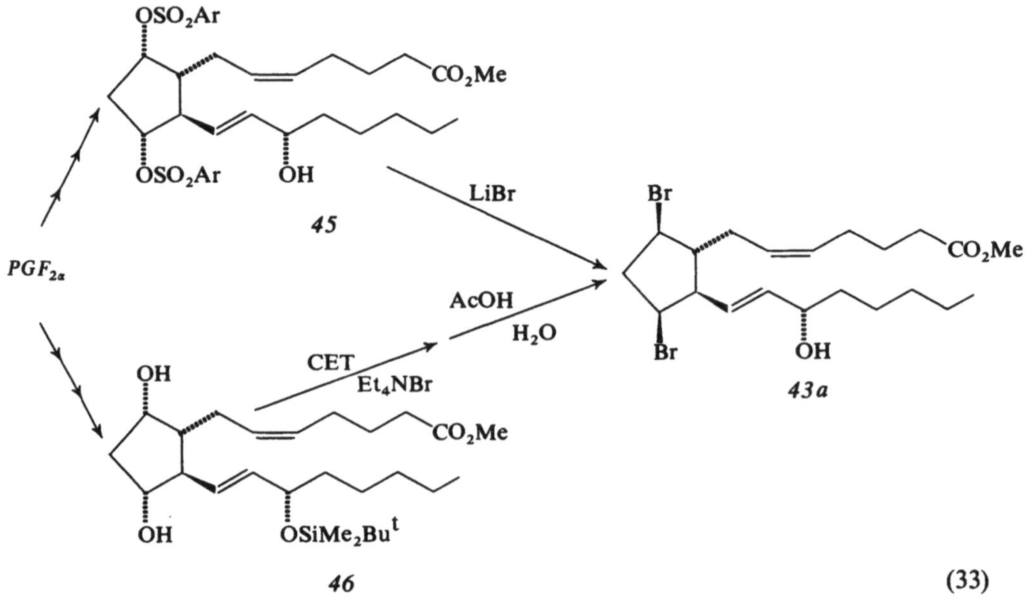

(33)

corresponding 9β, 11β-dibromide (50%) free of diastereoisomers, and thus avoided the tedious chromatographic purification required in the arenesulphonate route; hydrolysis then gave *43a*. With this modification, the overall yield for the conversion of $PGF_{2\alpha}$ into PGH_2 was raised to 7.0%.

The introduction of this mild alkyl halide synthesis also opened the way for the first chemical synthesis of the prostaglandin endoperoxide PGG_2 [47]. Reaction of *43a* with CET and benzyltributylammonium chloride provided the 15β-chloride *44a* in 82% yield (Eq. 32) and hydrolysis to the free acid *44b* was effected with hog pancreas lipase. In an elegant final step the 9,11-peroxide bridge and the 15-hydroperoxide group were simultaneously introduced via the silver trifluoroacetate-hydrogen peroxide method in a triple displacement reaction (Eq. 32). The yield was estimated at 15–20% based on biological assays and the PGG_2 was purified by HPLC (silica) at −5 °C.

As the first isolable intermediate in the bioconversion of arachidonic acid into prostaglandins and thromboxanes (Eq. 3), PGG_2 is a bicyclic peroxide of immense biological importance. It is difficult to obtain pure from natural sources and the presence of the 15-hydroperoxide group adds a further dimension of chemical lability to that associated with the 9,11-peroxide bridge. The chemical synthesis of PGG_2 is thus a landmark in prostaglandin chemistry. It also represents a pinnacle of success for the silver-salt route to bicyclic peroxides.

1.3 Peroxymercuration with reduction and brominolysis

In a process known as peroxymercuration, hydrogen peroxide and alkyl hydroperoxides can be alkylated by alkenes in the presence of a suitable mercury(II) salt

(Eq. 34). The resultant β-mercurioalkyl peroxides can often be demercurated with sodium borohydride (Eq. 35), or by brominolysis (Eq. 36) without substantial cleavage of the O-O bond. Both peroxymercuration and demercurations occur rapidly under mild conditions [48].

$$ROOH + R^1CH{=}CHR^2 + HgX_2 \rightarrow ROOCH(R^1)CH(R^2)HgX + HX \quad (34)$$

$$ROOCH(R^1)CH(R^2)HgX + NaBH_4 \rightarrow ROOCH(R^1)CH_2R^2 + Hg \quad (35)$$

$$ROOCH(R^1)CH(R^2)HgX + Br_2 \rightarrow ROOCH(R^1)CH(R^2)Br + HgXBr \quad (36)$$

In a variation of this method, some novel bicyclic peroxides have been prepared via the reaction of cyclic dienes with hydrogen peroxide and mercury(II) trifluoroacetate. Thus starting from *cis,cis*-1,5-cyclooctadiene, 2,6-bis(trifluoroacetoxymercurio)-9,10-dioxabicyclo[3.3.2]decane *47* was isolated in 42% yield [34,49]. A ^{13}C n.m.r. spectroscopic analysis of the crude product revealed that it contained an approximately equimolar amount of the corresponding bicyclic ether *48* and a smaller quantity of bismercurated 9-oxabicyclo[4.2.1]nonane *49*, but no [4.2.2]peroxide was detected (Eq. 37; $X = O_2CCF_3$). The organomercury trifluoroacetates were separated by dissolving the mixture in benzene, where upon crystals of the pure (solvated) peroxide *47* were precipitated.

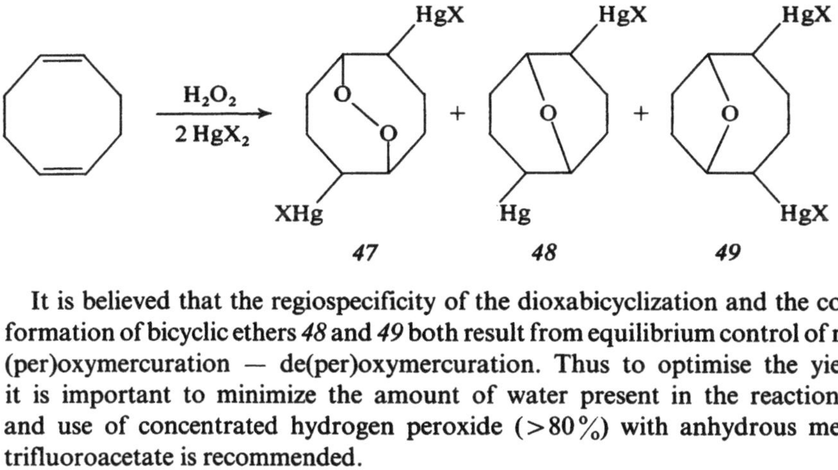

(37)

It is believed that the regiospecificity of the dioxabicyclization and the concurrent formation of bicyclic ethers *48* and *49* both result from equilibrium control of reversible (per)oxymercuration — de(per)oxymercuration. Thus to optimise the yield of *47* it is important to minimize the amount of water present in the reaction mixture and use of concentrated hydrogen peroxide (>80%) with anhydrous mercury(II) trifluoroacetate is recommended.

Attempts to use concentrated hydrogen peroxide with other mercury(II) salts for the dioxabicyclization of *cis,cis*-1,5-cyclooctadiene proved unsuccessful. With mercury(II) nitrate monohydrate, which is the reagent of choice for the analogous synthesis of monocyclic peroxides from acyclic dienes, [50,51] only [3.3.1]-ether *48* was obtained, and with mercury(II) acetate acetoxymercuration was the preferred reaction. An attempt to eliminate the concurrent formation of bicyclic ethers by engendering *kinetic* control through the use of a large excess of 30% H_2O_2 with mercury(II) acetate also failed. The formation of some [4.2.2]-peroxide in this reaction is mechanistically significant but of no synthetic value.

Borohydride reduction of the organomercury peroxide *47* yielded 9,10-dioxabicyclo-

[3.3.2]decane *50*, while brominolysis in dichloromethane gave a mixture of the three diastereoisomers of 2,6-dibromo-9,10-dioxabicyclo[3.3.2]decane *51* (Eq. 38) [34,49].

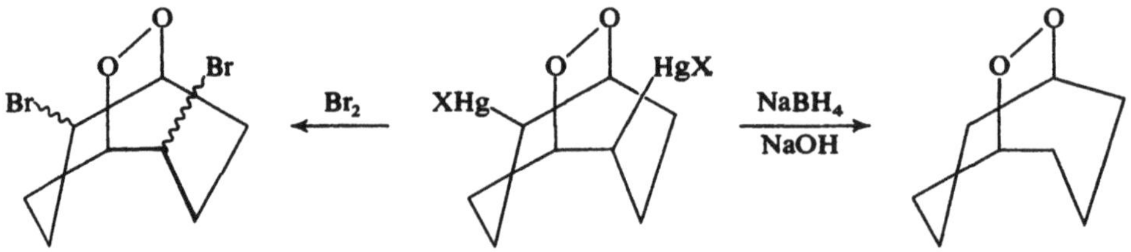

51 *47* *50*

(38)

The reductive demercuration was marred by the loss of about half of the peroxide due to competing deoxymercuration which afforded 4-cycloocten-1-ol. An additional complication was the formation of a small amount of *trans*-1,2-epoxy-*cis*-cyclooct-5-ene. The bicyclic peroxide *50* was readily separated from the unsaturated alcohol by silica chromatography, but complete removal of the epoxide was more difficult. Preservation of the peroxide linkage was markedly higher in the bromodemercuration. The diastereoisomeric dibromoperoxides *51* were separated by HPLC, although only one isomer was fully characterised.

Peroxymercuration and demercuration of 1,4-cyclooctadiene similarly afforded 8,9-dioxabicyclo[5.2.1]decane *52* (Eq. 39) and a 1:2:1 mixture of the three diastereoisomers of the corresponding 2,6-dibromo-compound *53* (Eq. 40) [49,52].

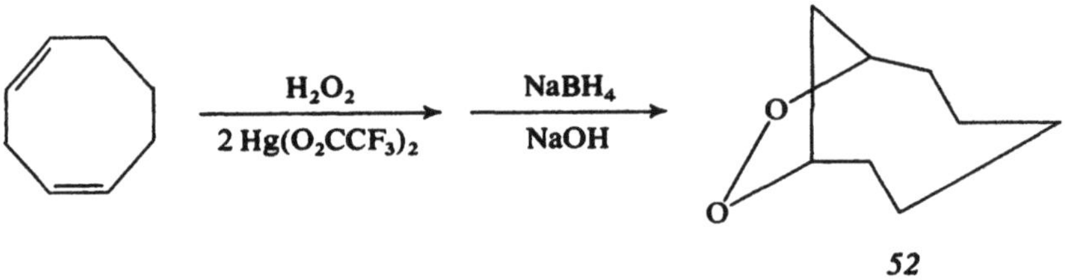

(39)

52

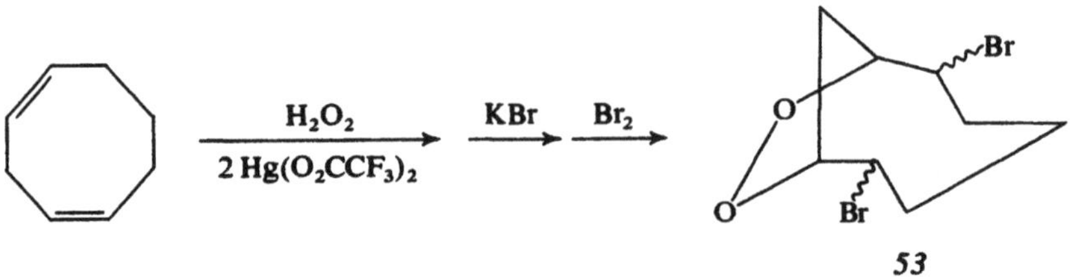

(40)

53

The reactions were cleaner than those with *cis,cis*-1,5-cyclooctadiene, and purification of the intermediate bis-mercurated peroxide was unnecessary. Indeed, it was a simple matter to isolate *52* (28 %) and *53* (38 %) from crude products obtained using a sample of 1,4-cyclooctadiene containing 30 % of *cis,cis*-1,5-cyclooctadiene, which is readily accessible via hydrobromination-dehydrobromination of the 1,5-diene. Two of the diastereoisomers of *53* were separately isolated by HPLC and the most abundant isomer in the mixture was unambiguously identified as that with one bromine *cis* and the other *trans* to the peroxide bridge.

An alternative entry into the [5.2.1] system was achieved by treating 2-cyclooctenyl hydroperoxide with mercury(II) trifluoroacetate then bromine (Eq. 41) [53].

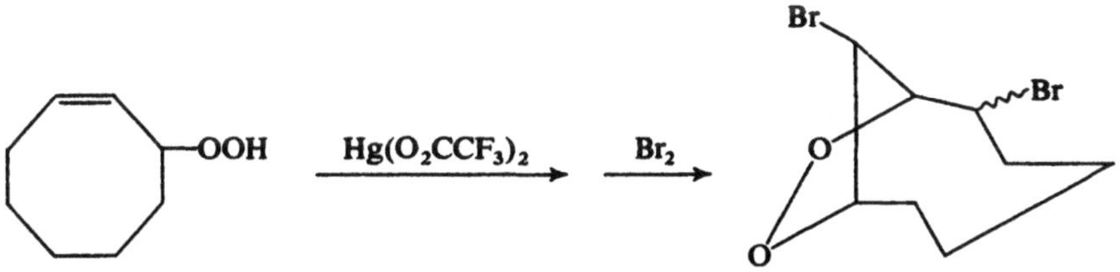

54

(41)

Formation of the intermediate organomercury peroxide *56* was rationalised in terms of an allylic mercuration providing an unsaturated hydroperoxide *55* that can cyclise by the favoured [54] 5-*exo* mode (equation 42; $X = O_2CCF_3$). However, this was not the main reaction pathway and the yields (2.7% and 0.6% of 2,*cis*-10-dibromo-8,9-dioxabicyclo[5.2.1]decanes *54* were an order of magnitude lower than those of the 2,6-dibromides *53* obtained from 1,4-cyclooctadiene.

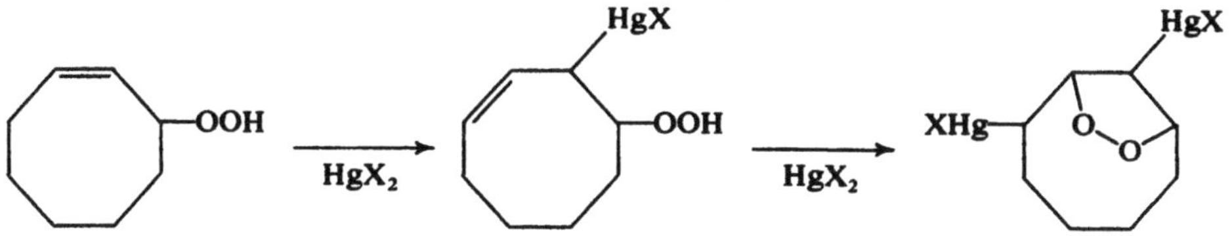

55 *56*

(42)

Our studies suggest that the peroxymercuration strategy is not well suited to the construction of smaller bicyclic peroxide skeletons. Thus, no evidence could be obtained for the formation of [2.2.1]-peroxides from reactions with several 5,5-disubstituted cyclopentadienes, [49] and even 3-cyclopentenyl hydroperoxide resisted intramolecular peroxymercuration, molecular addition of mercury(II) trifluoroacetate apparently being preferred [41]. The [3.2.1]-system was generated via peroxymercuration of 1,4-cyclohexadiene (Eq. 43), but the yield was low [49].

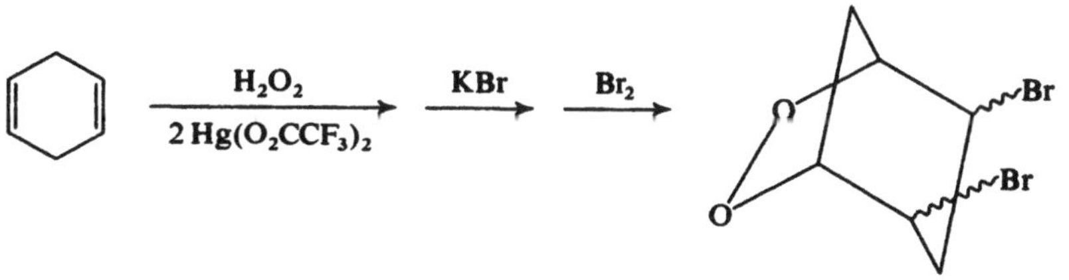

57

(43)

A bicyclic peroxide was isolated in 1.8% yield by HPLC of the bromodemercuration product, and was identified as a single diastereoisomer of 2,4-dibromo-6,7-dioxa-bicyclo[3.2.1]octane *57* with the *trans,trans-* or *cis,cis*-configuration. By analogy with the cyclooctadiene reactions, formation of the other two diastereoisomers of *57* can be expected, but although additional peroxides with similar HPLC characteristics were detected, they were not identified. Thus the presence of [2.2.2]-compounds cannot be ruled out, and no comment can be made on the regioselectivity of the dioxabicyclization.

Thus the range of bicyclic peroxides available via peroxymercuration may be quite limited. Nevertheless where the method works best, namely with 1,5- and 1,4-cyclo-octadiene, it makes a valuable contribution in that each peroxymercuration is regiospecific and leads to a dioxabicyclodecane that is isomeric with the [4.2.2] compound *23* available via photooxygenation (Eq. 18). Furthermore, the [3.3.2] compounds derived from 1,5-cyclooctadiene are, to the best of our knowledge, the only bicyclic peroxides obtained to date that do not contain either a 5- or a 6-membered dioxacycloalkane ring.

1.4 Triflate and diradical routes to 2,3-dioxabicyclo[2.2.1]heptane 9.

To add to the silver salt (Eq. 19) and peroxymercuration (Eq. 34–36) methods of preparing dialkyl peroxides, a third mild alkylation procedure has been developed that involves the use of alkyl trifluoromethane sulphonates (triflates) [55]. The peroxide transfer reaction between bistributyltin peroxide and the bistriflate *58* of *cis*-1,3-cyclopentanediol provided one of the first syntheses of 2,3-dioxabicyclo[2.2.1]heptane *9* (Eq. 44, H for D; Tf = O_2SCF_3) [56]. Because of the sensitivity of *9*, it was necessary to carry out the reaction *in vacuo* with rapid transfer of the volatile products to a cold trap to avoid decomposition; a yield of 22% was achieved.

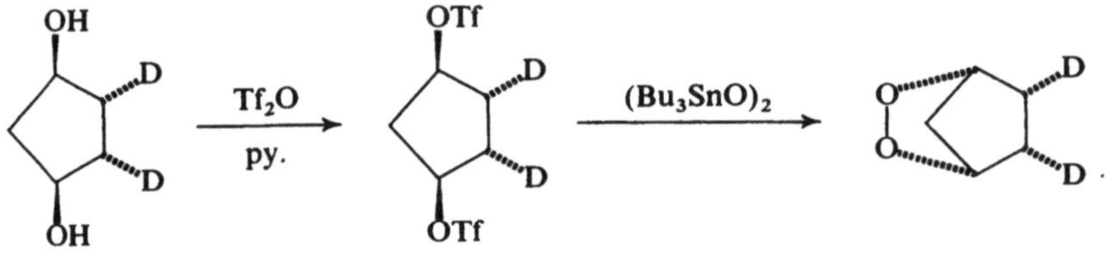

58

(44)

Since the bistriflate *58* can be prepared from the diol by reaction with triflic anhydride and pyridine, the overall conversion here is the same as that effected by Ph_3PBr_2 then H_2O_2/AgO_2CCF_3 (Eq. 31). However, there is the important difference that the transformation from diol to peroxide now proceeds with net predominant inversion, as shown by reaction of the *cis*-4,5-dideuterio compound (Eq. 44). The method would not, therefore, be suitable for converting an F prostaglandin into the natural endoperoxide. Although a potentially general method, no other bicyclic systems have yet been prepared by it. Where the stereochemistry of diol-peroxide conversion is unimportant, it seems likely that the alternative silver salt procedure will be preferred on the grounds of experimental convenience.

In addition to the photoxygenation/diimide (equation 6), [16] silver salt (Eq. 22), [36] and triflate (Eq. 44) [56] routes, *9* has also been prepared by benzophenone-sensitized photodecomposition of the corresponding azo compound *59* and trapping of the resultant triplet diradical with oxygen (Eq. 45) [57].

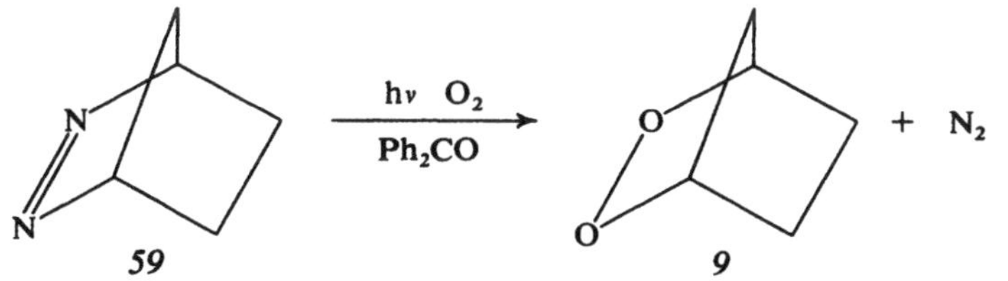

59 *9* (45)

Correct experimental conditions are vital. In particular, it is essential to irradiate only the benzophenone chromophore, which can be achieved by employing an appropriate u.v. laser, for direct excitation of the azo compound produces a singlet diradical that collapses to bicyclopentane. Oxygen pressure (150 psi) and reaction time (60–70 h) must be carefully regulated to obtain optimum yields (ca. 20%) of *9*.

Although no other examples have been reported, oxygen trapping of azo-derived triplet diradicals provides a potentially versatile strategy for the synthesis of bicyclic peroxides under neutral conditions.

2 Spectroscopy

The structures of the new bicyclic peroxides have been established by the usual combination of physical techniques and chemical transformations. Here we highlight features of the ^1H and ^{13}C n.m.r. spectroscopic data that provide the best characterization of these compounds; their reactions are discussed later. Information about the C-O-O-C dihedral angle in organic peroxides is potentially available from photoelectron (PE) spectroscopy. Measurements on comparatively rigid systems play an important part in establishing a soundly based experimental correlation, and the results obtained on several of these bicyclic peroxides are presented in this section also.

2.1 N.M.R. Spectra

The most distinctive features of the ^1H and ^{13}C n.m.r. spectra of bicyclic peroxides are provided by their bridgehead nuclei. An analysis of the data on over thirty compounds indicates that the characteristic chemical shift ranges are δ 3.7–4.8 for bridgehead protons and δ 72–89 for bridgehead carbons.

^{1}H.N.M.R. In the [2.2.1] systems the bridgehead protons appear in the range δ 4.4–4.8, usually in the form of a broad singlet. The chemical shift of these protons is only slightly affected by the presence of bromine, alkylidene, or oxo substituents at the 7-position, or by *endo*- and *exo*-bromine and *exo*-trifluoroacetate groups at C-5. However, a 7-cyclopropyl group leads to an upfield shift (δ 3.80) and, not surprisingly, the 7-oxa-compounds (ozonides) have resonances at lower field

Table 1. Proton-decoupled ^{13}C N.M.R. Spectra of Bicyclic Peroxides[a,b]

[2.2.1]

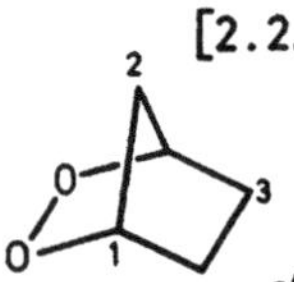

9[c]

C	δ
1	78.69
2	43.80
3	29.15
Ref	58

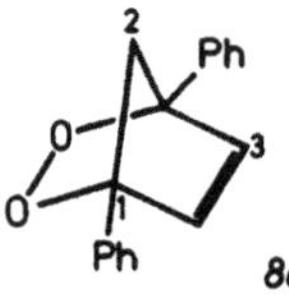

8a[d]

C	δ
1	95.15
2	61.12
3	137.90
Ref	26

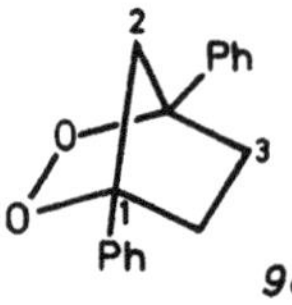

9a[e]

C	δ
1	91.25
2	52.60
3	35.21
Ref	28

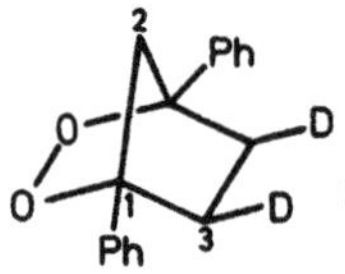

C	δ
1	91.29
2	52.96
3	35.10
Ref	28

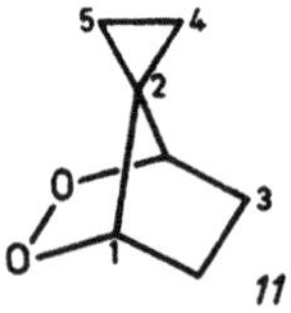

11

C	δ
1	82.98
2	39.66
3	29.70
4	6.33
5	3.25
Ref	58

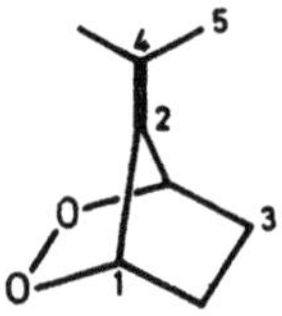

10

C	δ
1	75.76
2	138.89
3	29.37
4	121.12
5	21.51
Ref	58

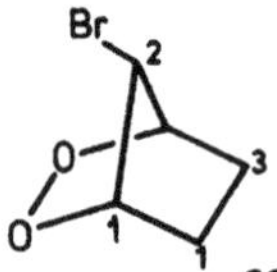

26

C	δ
1	82.36
2	55.12
3	27.85
Ref	38

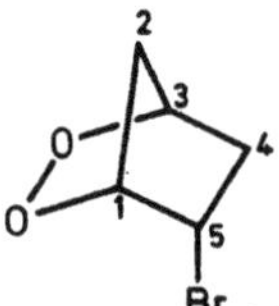

33

C	δ
1	81.24
2	43.67
3	78.74
4	40.74[k]
5	47.13[k]
Ref	40

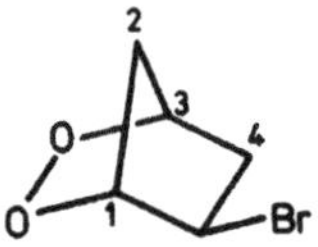

34

C	δ
1	81.67
2	40.66
3	78.28
4	42.63
5	43.81[k]
Ref	40

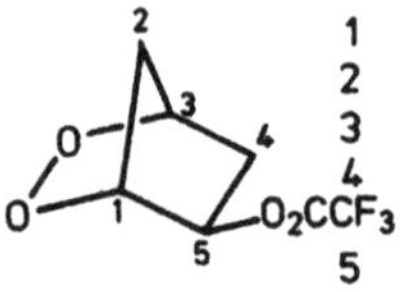

35

C	δ
1	77.45
2	41.19
3	76.77
4	38.19
5	74.94
Ref	40

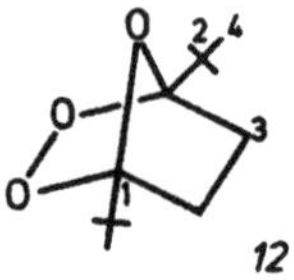

12

C	δ
1	116.44
2	33.24
3	29.83
4	25.77
Ref	58

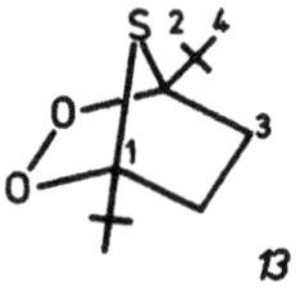

13

C	δ
1	114.66
2	35.24
3	35.24
4	28.02
Ref	58

[3.2.1]

27

C	δ
1	82.89
2	60.39
3	32.07
4	16.06
Ref	38

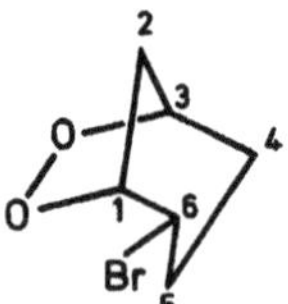

40

C	δ
1	81.19
2	47.11
3	74.85
4	31.59
5	30.02
6	50.92[k]
Ref	41

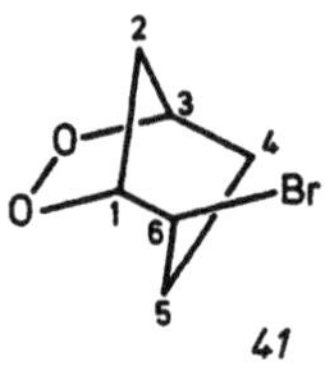

41

C	δ
1	79.15
2	42.77
3	76.17
4	27.78
5	26.95
6	48.09[k]
Ref	41

57[i]

C	δ
1	76.48
2	43.04
3	38.14
4	20.91
Ref	49

[2.2.2]

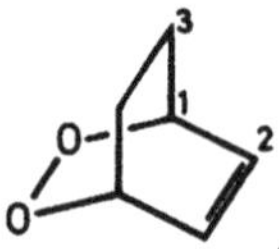

2-H_2

C	δ
1	70.85
2	132.26
3	21.69
Ref	58

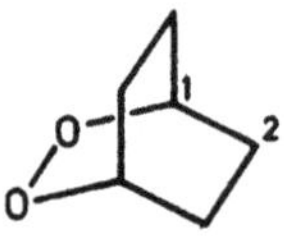

2

C	δ
1	71.70
2	24.29
Ref	58

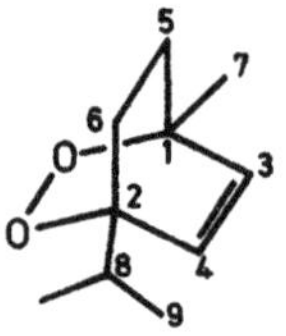

1[f]

C	δ
1	74.32
2	79.77
3	133.06
4	136.42
5	25.63
6	29.54
Ref	58

18-H_2

C	δ
1	73.70
2	126.57
3	5.98
4	4.30
Ref	30d

[4.2.1]

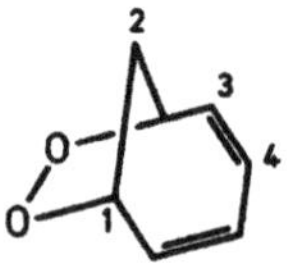

22-H_4

C	δ
1	73.59
2	38.92
3	128.52
4	132.15
Ref	30d

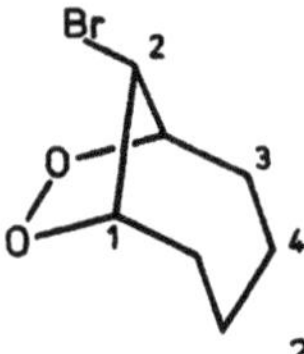

28

C	δ
1	87.50
2	57.40
3	32.57
4	22.63
Ref	38

Table 1 (continued)

[3.2.2]

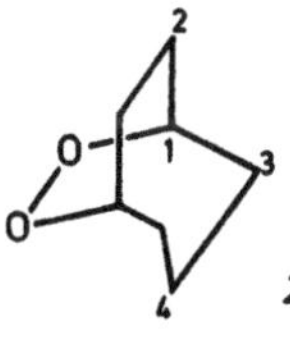

20

C	δ
1	76.84
2	20.43
3	34.95
4	19.89
Ref	28

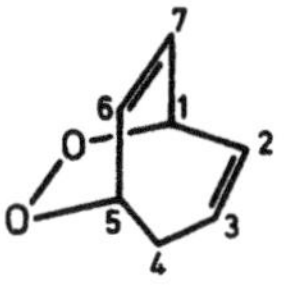

20-H_4

C	δ
1	75.59
2	130.58
3	134.04
4	35.67
5	73.42
6	125.00
7	128.84
Ref	30d

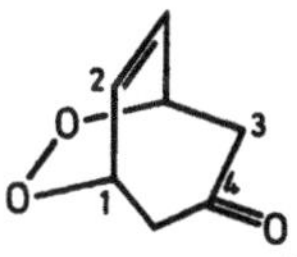

21-H_2

C	δ
1	71.96
2	131.43
3	50.43
4	192.81
Ref	58

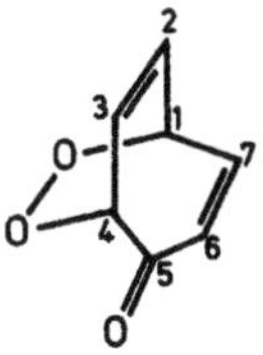

C	δ
1	75.55
2	123.37
3	131.54
4	85.78
5	195.13
6	144.78
7	138.25
Ref	58

[5.2.1]

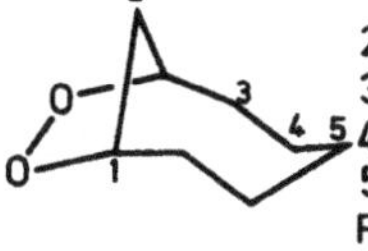

52

C	δ
1	77.61
2	45.18
3	33.12
4	25.13
5	26.14
Ref	52

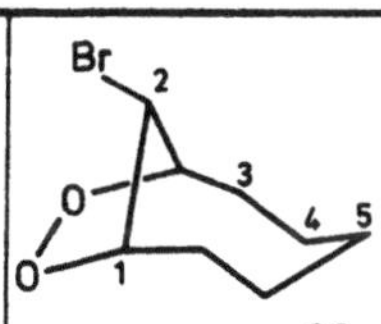

29

C	δ
1	88.59
2	62.04
3	32.28
4	25.83
5	25.31
Ref	38

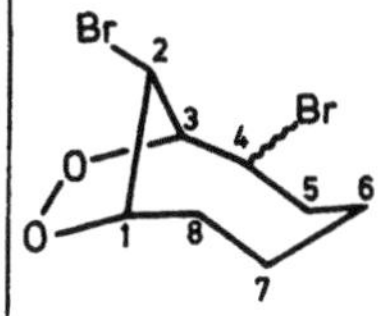

54[g]

C	δ
1	88.88
2	58.76
3	93.34
4	52.50
5	37.68
6	25.33
7	26.33
8	30.66
Ref	53

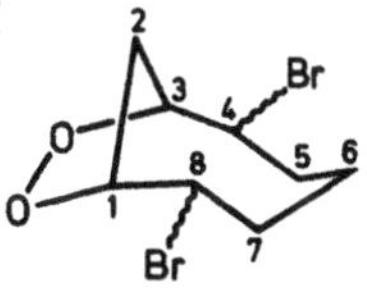

53[h]

C	δ
1	83.30
2	42.61
3	81.26
4	53.79
5	37.32
6	21.59
7	34.60
8	53.55
Ref	49

53[i]

C	δ
1/3	80.66
2	40.00
4/8	52.17
5/7	34.60
6	16.39
Ref	49

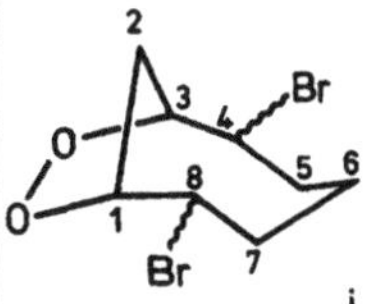

53[i]

C	δ
1/3	83.58
2	42.50
4/8	54.79
5/7	37.29
6	17.12
Ref	49

[4.2.2]

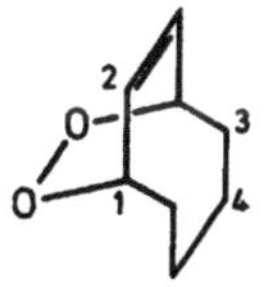

23-H_2

C	δ
1	76.58
2	128.24
3	33.25
4	24.08
Ref	58

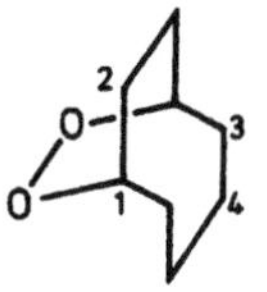

23

C	δ
1	76.02
2	24.46
3	34.60
4	20.68
Ref	34

[3.3.2]

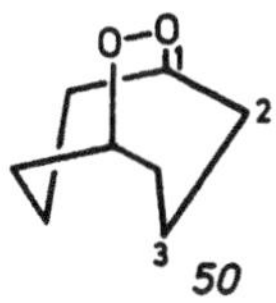

50

C	δ
1	83.96
2	31.38
3	23.82
Ref	34

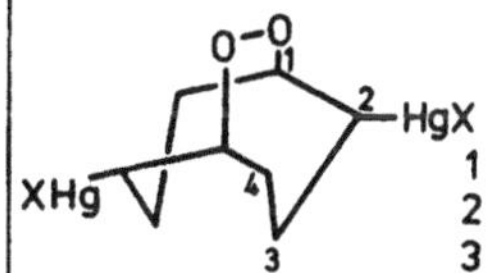

47[j]

C	δ
1	85.27
2	48.42
3	34.06
4	29.09
Ref	49

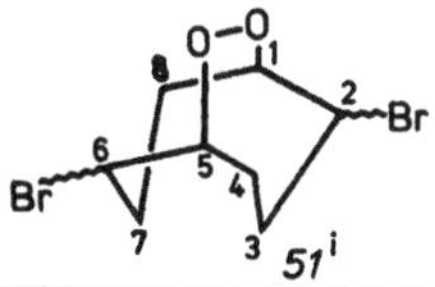

51[i]

C	δ
1/5	86.15
2/6	48.18
3/7	32.75
4/8	26.23
Ref	49

[a] δ (p.p.m.) downfield from T.M.S. for solutions in $CDCl_3$. [b] Assignments are uncertain where chemical shifts of carbons in equal abundance are very similar. [c] Lit.[36] δ 78.8, 43.8, 29.1; lit.[28] δ 78.7, 43.8, 29.1. [d] Phenyl carbons: δ 126.9, 128.8, 129.13, 133.36. [e] Phenyl carbons: δ 126.8, 127.7, 128.6, 134.9. [f] Other carbons: δ 21.40 (7), 32.17 (8), 17.21 (9). [g] Stereochemistry at C-4 uncertain. [h] *cis, trans.* [i] *cis, cis* or *trans, trans.* [j] *trans, trans*; X = O_2CCF_3; PhH solvate in $CH_2Cl_2/CDCl_3$. [k] Assignment confirmed by off-resonance decoupling.

(δ 5.75). For the parent 2,3-dioxabicyclo[2.2.1]heptane *9* and its 1,4-diphenyl derivative *9a*, the multiplets for the remaining protons do not overlap and have been assigned on the basis of coupling patterns and comparison with the corresponding *cis,exo*-5,6-dideuterio derivatives [28].

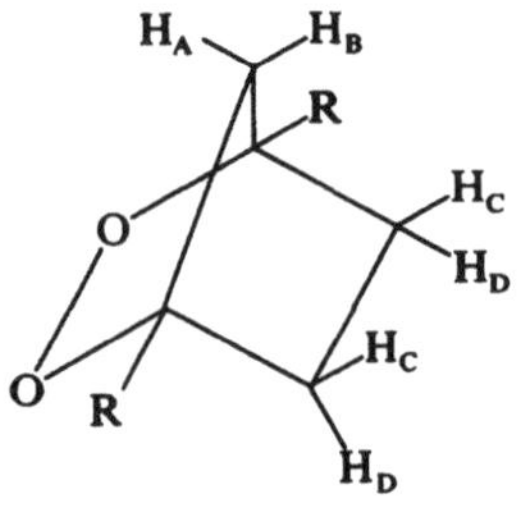

9, R=H
9a, R=Ph

On the methylene bridge of *9* the proton *cis* to the peroxide linkage (H_A) appears at δ 1.87 (solvent PhH) and shows long range W-plan coupling with the 5-, and 6-*endo*-protons (H_D), while the corresponding *trans*-proton (H_B) exhibits no 4-bond coupling and is at δ 1.35. Compound *9a* shows a similar spectrum but offset to lower field by about 0.8 ppm [28].

For a wide variety of higher [n.2.1] systems, the chemical shift of the bridgehead protons is in the range δ 4.1–4.8 and thus closely resembles that in the [2.2.1] compounds; the presence of unsaturation in the bridges results in only a small downfield shift (0.1–0.2 ppm). For several higher [n.2.2] compounds on the other hand, the bridgehead protons resonate at δ 3.7–4.5, a little upfield of those of the [2.2.1] peroxides; unsaturated bridges again produce a downfield shift which is now usually of the order of 0.5 ppm.

^{13}C.N.M.R. In Table 1 we present what, at the time of writing (December 1980), is a comprehensive set of available ^{13}C n.m.r. data on both saturated and unsaturated bicyclic peroxides; many previously unpublished results are included. Of the eight skeletal types known, the [4.2.1]-peroxide *22* and the [3.2.1]-peroxide (unknown) are the only unsubstituted members for which no data are presently available. Most of the unsubstituted peroxides have bridgehead carbon resonances in the narrow range δ 77–79, only the [2.2.2]-compound *2* (δ 72) and the [3.3.2]-compound *50* (δ 84) being markedly different. The introduction of unsaturated bridges, bromine and/or various other groups (see Table) extends the range of observed chemical shifts for bridgehead carbons, but only by about 10 ppm, at the higher end. Only when bridgehead phenyl substituents or additional heteroatoms are present in the [2.2.1] system are shifts outside the δ 72–89 range observed.

Bridgehead ^{13}C-^{1}H coupling constants have been measured [28] for the series of [n.2.2] homologues *9* (n = 1), *2* (n = 2), *20* (n = 3), and *23* (n = 4). Compounds *23* and *20* were found to have values (141 Hz and 144 Hz respectively) similar to that (141 Hz) found for the α-C-H coupling constant in diisopropyl peroxide, but compounds *2* and *9* showed significantly higher values (150 Hz and 161 Hz respectively) in keeping with the greater ring strain in these peroxides.

2.2 PE Spectra

The He(I) photoelectron spectra of compounds *9, 10, 11, 12, 2, 2-H_2, 23, 23-H_2*, and *50*[59] and of *26–29*[60] have been recorded. The spectra of peroxides *9, 2*, and *23* have been measured independently by another group and the results are in good agreement with ours; data for compounds *6, 2a*, and *20* were also reported[28].

Interest lies in the separation between the two highest occupied molecular orbitals associated with the peroxidic oxygens since calculations predict a dependence of the energy difference between these orbitals upon the C-O-O-C dihedral angle, θ. Hence bands in each PE spectrum were assigned to ionisations from these orbitals and an observed energy difference (ΔI) thereby determined. In the parent skeletons *9, 2, 23*, and *50*, the ionisations of interest give rise to the first two peaks of each spectrum, but in the other compounds bands assigned to ionisation events from ethylenic π-orbitals (in *10, 2-H_2, 23-H_2*). Walsh orbitals (in *11*), or lone pairs (in *12, 26–29*) come between them. Approximate values of the dihedral angles θ in the bicyclic peroxides can be estimated from Dreiding models, but the precise values are unknown. However, the geometries were optimised within the theoretical framework used.

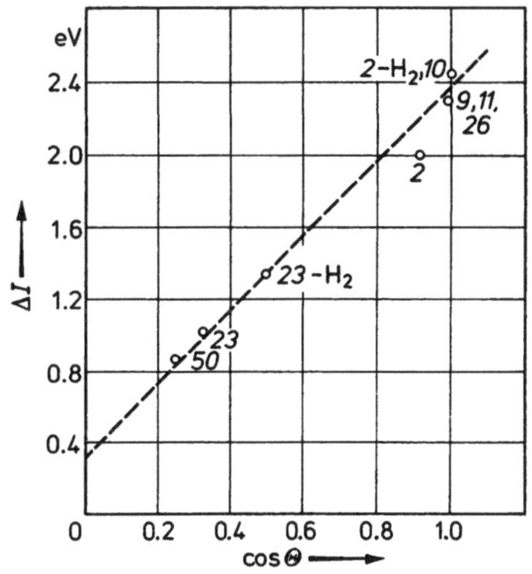

Diagram 1. Correlation between the split (ΔI) of the first two ionisation potentials associated with the peroxide moiety and the C—O—O—C dihedral angle (cos θ) for selected bicyclic peroxides.

A correlation between ΔI and cos θ was found and is shown in diagram 1. Although this is in line with theoretical calculations, others[28] have correlated θ itself rather than cos θ with ΔI. The fact that both correlations seem to be valid suggests that PE spectra cannot yet be used with confidence to determine dihedral angles in peroxides. More points are needed and, of course, experimental measurement of dihedral angles in judiciously chosen systems would be a great step forward.

3 Reactions

As described at the beginning of this review, 2,3-dioxabicyclo[2.2.1]heptane derivatives *4* have been established as the biological precursors of prostaglandins, thromboxane A_2 and prostacyclin (Eq. 2). Analogous chemical transformations were therefore expected for the prostanoid bicyclic peroxide model compounds that have been synthesised during the last four years. Indeed, reductions mimicking the formation of the *cis*-1,3-cyclopentanediol product $PGF_{2\alpha}$ and isomerisations parallel to the production of the hydroxyketone PGE_2 have been demonstrated and will now be described. However, to the best of our knowledge no simple bicyclic peroxides have yet shown reactions analogous to the formation of thromboxane A_2 or prostacyclin.

3.1 Reductions

The conversion of the parent bicyclo[2.2.1]-system *9* into *cis*-1,3-cyclopentanediol *60a* (Eq. 46) "connects" directly with the biogenesis of $PGF_{2\alpha}$ from the natural endoperoxide *4* through the action of a reductase (Eq. 2).

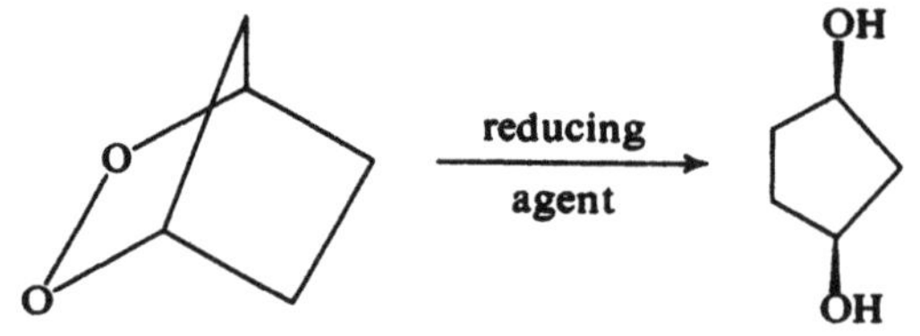

(46)

9 *60a*

A number of reducing agents have been employed for this purpose. For example, catalytic hydrogenation (Pd-C) [16, 56] and thiourea reduction [36] gave the expected *60a* in high yield. In the latter case *60a* was not directly characterized, but was converted to the *cis*-diacetate derivative. Similarly, the *cis*-cycloalkanediols *60b–i* were prepared, most of them quantitatively, from the corresponding bicyclic peroxide. In fact, these reductions proved helpful in the chemical characterization of the bicyclic peroxides, and in establishing the configurations of the [2.2.1]-peroxides that afford *60d–f* [40]. Catalytic hydrogenation is more convenient in view of the simplicity of the work-up; however, for *60c* thiourea reduction was essential to preserve the exocyclic double bond. In the case of *60b* the peroxide bond is selectively reduced on catalytic hydrogenation without severing the cyclopropane ring.

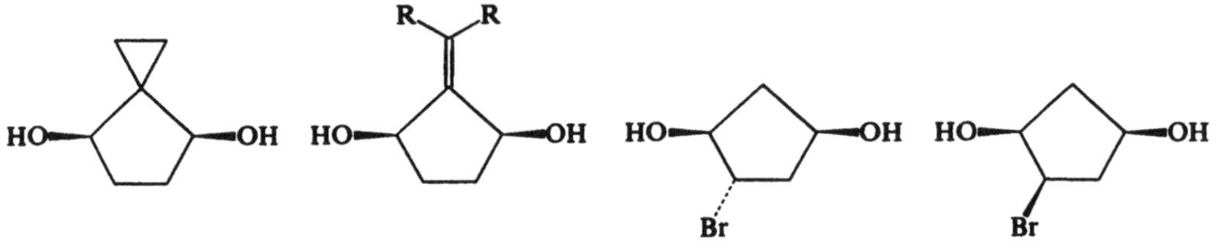

60b [21] *60c* [20] *60d* [40] *60e* [40]

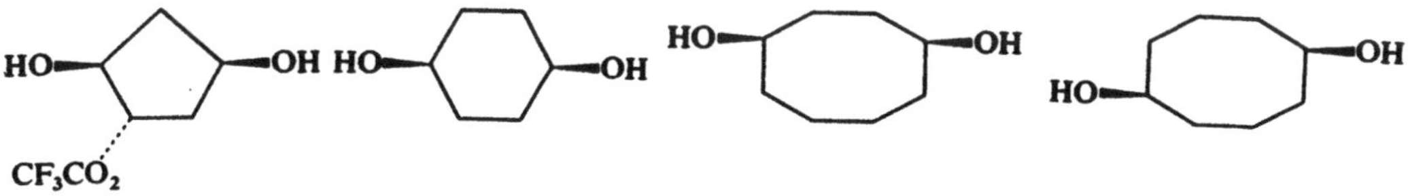

60f[40] *60g*[27] *60h*[34] *60i*[34]

The bicyclic ozonides *12*[23)] and thiaozonides *13*[25)] afford on catalytic hydrogenation (Pd-C) the expected 1,4-diones *61* (Eq. 47). Alternatively, deoxygenation of *12* or desulfurization of *13* with triphenylphosphine led to the same products essentially quantitatively. Both reductions served for the chemical characterization of these bicyclic peroxides.

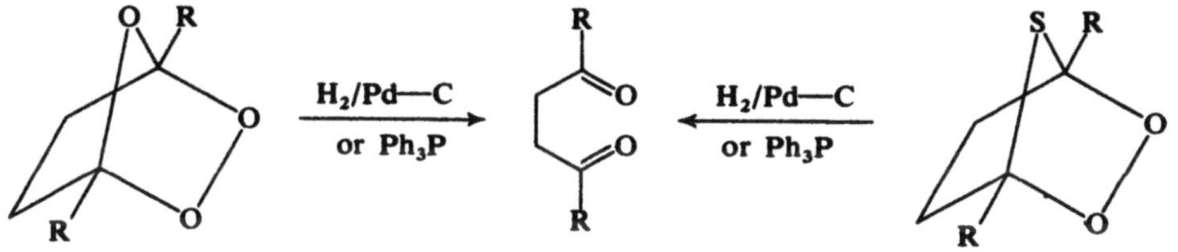

12 (R = H, Me, ${}^tBu$, Ph) *61* *13* (R = Me, ${}^tBu$)

(47)

3.2 Base-Catalysed Isomerizations

The well established chemistry of acyclic secondary-alkyl peroxides [12)] suggested that bases should catalyse the isomerization of related bicyclic peroxides to cyclic hydroxyketones *62* via abstraction of bridgehead hydrogen and heterolysis of the peroxide bond (Eq. 48).

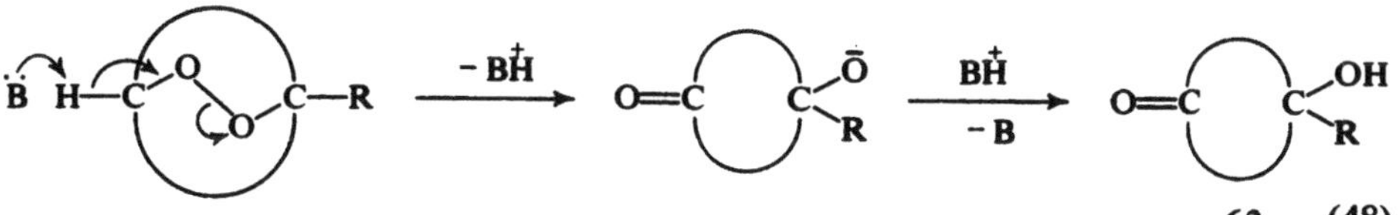

62 (48)

This prediction was borne out for reactions of 2,3-dioxabicyclo[2.2.2]octane *2*[27)] and the 7-substituted-[2.2.1]-peroxides *10*[20)] and *11*[21)] which afforded *62b–d* respectively upon treatment with either methanolic KOH or triethylamine. The bismercurated-[3.3.2]-peroxide *47* reacted similarly with aqueous NaOH, but the resultant 5-hydroxycyclooctanone underwent further rearrangement to the internal hemi-ketal.[49)]

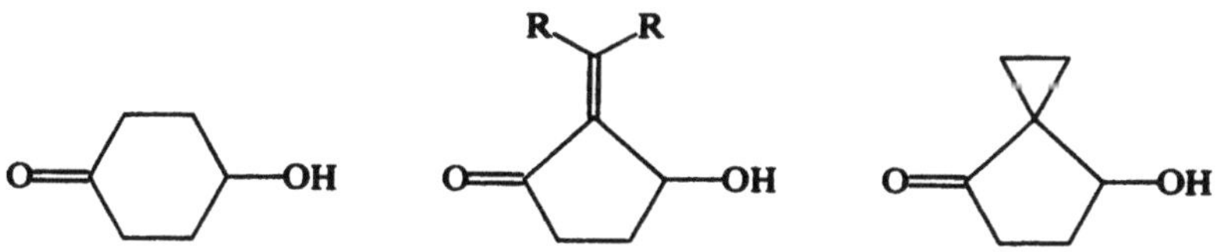

62b[27] *62c*[20] *62d*[21]

Particular interest surrounds the reaction of 2,3-dioxabicyclo[2.2.1]heptane *9* with bases since the expected isomerization into 3-hydroxycyclopentanone *62a* would "connect" directly with the conversion of prostaglandin endoperoxide *4* into PGE_2 (Eq. 2). However, substantial quantities of levulinaldehyde *63*, a product with no analogue in the biosynthesis of prostaglandins, was obtained in addition to *62a* upon treatment of *9* with 1,4-diazabicyclo[2.2.2]octane (DABCO), [61] imidazole, [61] or methanolic KOH [16]. Kinetic studies of the DABCO-catalysed reaction indicated that fragmentation to *63* and rearrangement to *62a* are closely related mechanistically, and involve rate-determining cleavage of a bridgehead C-H bond [61]. It was suggested that the intermediate keto-alkoxide partitions between protonation to give hydroxy-ketone *62a* and retro-aldol cleavage leading to levulinaldehyde *63* (Eq. 49). The fact that authentic hydroxyketone *62a* does not undergo base-catalysed, retro-aldol cleavage into levulinaldehyde indicates that further investigation of the mechanistic origin of this unusual product is warranted.

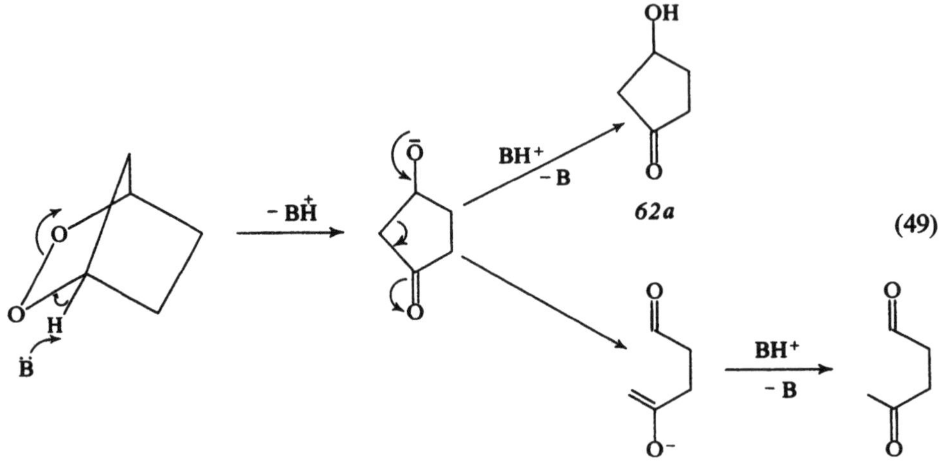

62a

63

(49)

An important observation from the viewpoint of prostaglandin endoperoxide chemistry was the fact that 2,3-dioxabicyclo[2.2.1]heptane *9* decomposes rapidly in *neutral* water, once again to give a mixture of *63* and *62a* [62]. It appears that even water is basic enough to abstract bridgehead hydrogens from *9*, emphasizing the extraordinary reactivity of the [2.2.1] system.

It should be pointed out that the mechanism shown in Eq. 49 is not the only

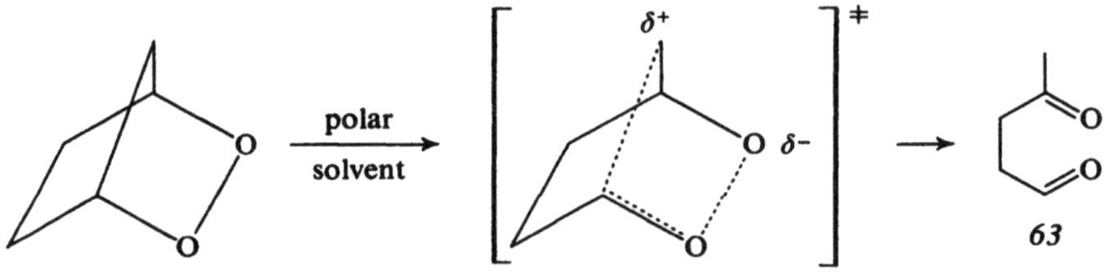

9

63

(50)

pathway by which *9* can be converted into *63*. Thus thermolysis of *9* gave mainly a mixture of *63* and 4,5-epoxypentanal *64*, the product distribution being strongly solvent dependent [62]. Formation of *64* is accounted for by the expected homolytic pathway (Eq. 52, discussed in the next sect.), while production of *63*, the proportion of which increased markedly with increasing solvent polarity, has been acribed to the mechanism shown in Eq. 50. It was suggested that the exceptionally high rate of formation of *63* observed in acetic acid arises through proton donation to the electron rich oxygen in the dipolar transition state.

3.3 Thermal Decompositions

Relatively little systematic work has been carried out on the thermal behaviour of the bicyclic peroxides compared to monocyclic systems [63]. From the few results obtained to date, oxygen diradicals appear to be reasonable intermediates to account for the observed products (Eq. 51).

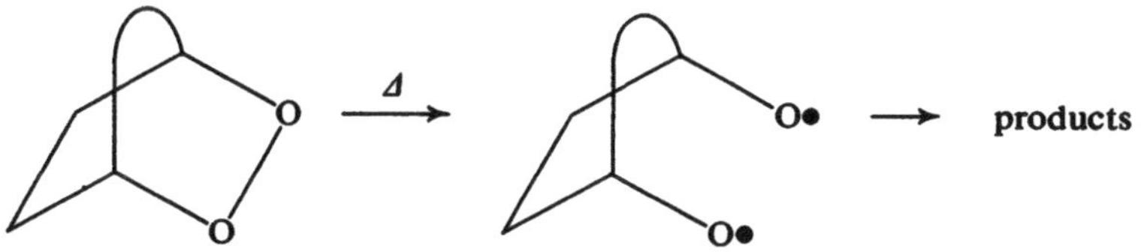

(51)

For example, the parent bicyclo[2.2.1] system *9* affords the epoxy-aldehyde *64* in high yield (97%) in nonpolar solvents (cyclohexane) [62]. The mechanism is rationalized in equation 52. Small amounts (ca. 1%) of the bicyclic ether *65* are also formed, but it is known that *64* rearranges into *65* on heating [62].

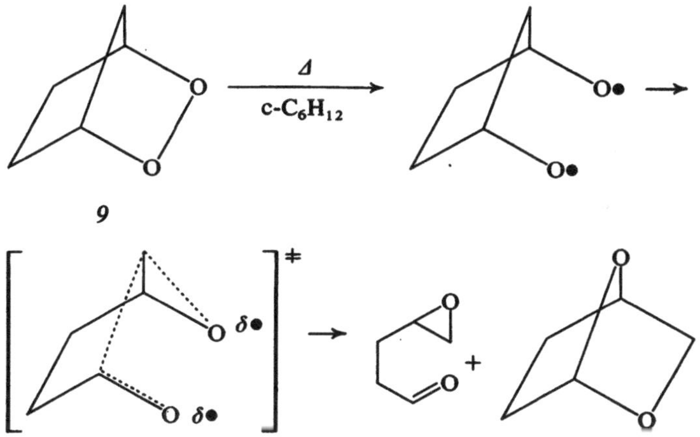

9 *64* *65*

(52)

To the best of our knowledge, no direct "connection" exists between this thermal behaviour of the model bicyclo[2.2.1] system *9* and prostaglandin endoperoxide *4*, since no natural epoxyaldehydes analogous to *64* have been reported as metabolites of *4*.

On the other hand, the 1,4-diphenyl derivative *9a* leads to the 1,3-dione *66* and ethylene (Eq. 53), but in very low yield (ca. 10%) [15].

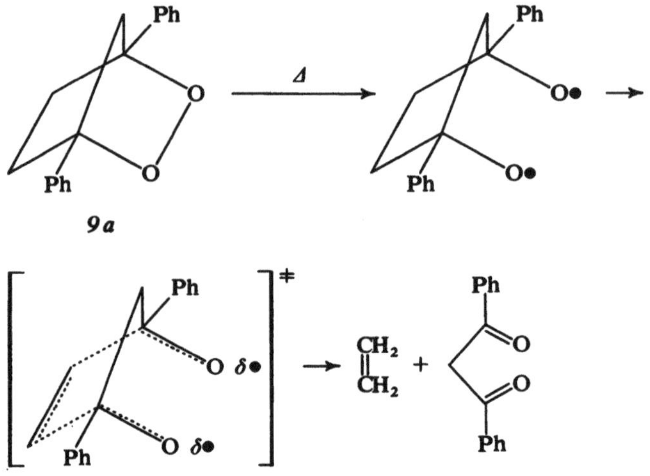

9a

66

(53)

Thus, whereas the thermal behaviour of the parent system *9* is akin to that of 1,2-dioxolanes, [64] the thermolysis of the 1,4-diphenyl derivative *9a* resembles that of 1,2-dioxanes [65]. Although the bicyclo[2.2.1]skeleton possesses both the 1,2-dioxolane and 1,2-dioxane rings, the divergent thermal behaviour between the parent and 1,4-diphenyl systems is difficult to rationalize. The poor product balance in the case of *9a* should be kept in mind!

In this context it must be noted that the natural prostaglandin endoperoxide *4* does give rise to malonaldehyde and 12-hydroxy-5,8,10-heptadecatrienoic acid *67* under biological conditions[66] (Eq. 54), which "connects" with the chemistry of the 1,4-diphenyl derivative *9a* and bespeaks a preference for 1,2-dioxane-type [65] rather than 1,2-dioxolane-type [64] fragmentation.

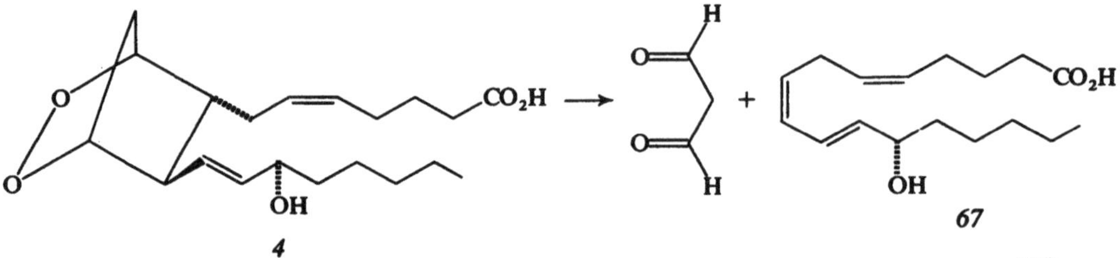

4

67

(54)

The bicyclo[2.2.2] system *2*, as expected, [63,65] leads on thermolysis (Eq. 55), to ethylene (23–100%) and succinaldehyde, but the latter deteriorates extensively at the high temperatures involved [67].

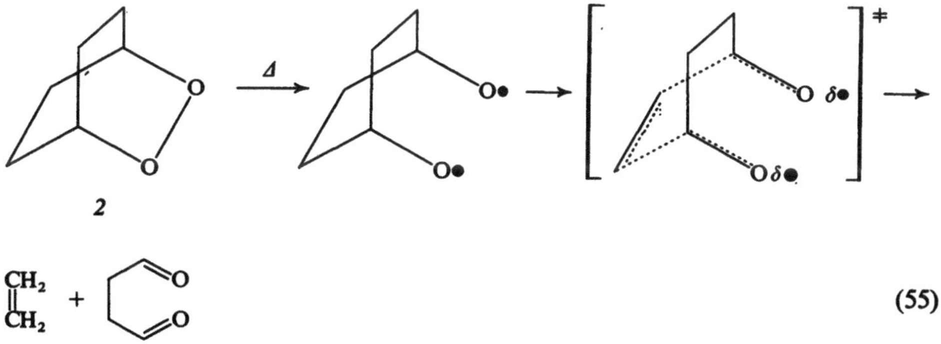

2

(55)

The mechanistic dichotomy (Eq. 50 and 52) that obtains in the thermolysis of 2,3-dioxabicyclo[2.2.1]heptane *9* has no equivalent with *2* and accordingly the abnormally large solvent effects found in the thermolysis of *9* were not observed for *2*. In fact, the [2.2.2] system 2 is considerably more stable thermally than the [2.2.1] system *9*. While *9* decomposes quite rapidly in cyclohexane at 60–80 °C, for *2* temperatures as high as 120–150 °C are needed to promote comparable rates; the activation parameters are $\Delta H^{\neq} = 21$ kcal mol^{-1} and $\Delta S^{\neq} = -19$ e.u. for 2,3-dioxabicyclo[2.2.1]-heptane, and $\Delta H^{\neq} = 33$ kcal mol^{-1} and $\Delta S^{\neq} = +3$ e.u. for 2,3-dioxabicyclo[2.2.2]-octane [67].

Finally, it is of interest to mention that the lactone-peroxide *17* decarboxylates at ca. 140 °C to afford succinaldehyde, but with light emission (Eq. 56) [20].

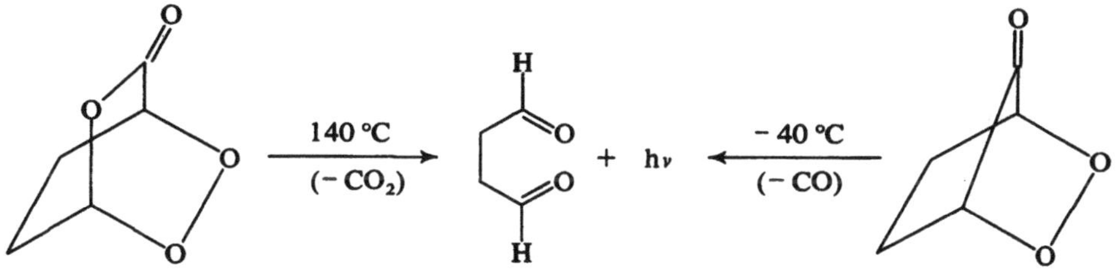

17 *15*

(56)

Similarly, the very labile keto-peroxide *15* also affords succinaldehyde with chemiluminescence on decarbonylation [20]. Sufficient chemical energy is stored in these bicyclic peroxides to generate electronically excited products [68].

4 Concluding Remarks

Studies on the biosynthesis of prostaglandins revealed that a strained bicyclic bis(secondary-alkyl) peroxide was not only a key intermediate, but could survive the biological conditions long enough to be isolated, albeit in minute amounts. This was a startling discovery in that no simple bicyclic peroxides of the same type were known, clearly for reasons of preparative difficulty. The "prostaglandin connection" focused attention upon this missing class of peroxides and stimulated efforts to fill the gap. A principal aim of this review has been to show how this synthetic challenge was met successfully.

Use of mild conditions was crucial and the development of diimide reduction of singlet oxygenates, silver-salt-assisted displacement of halide by peroxide nucleophiles, peroxymercuration and demercuration, peroxide transfer from organotin to alkyl triflates, and oxygen trapping of azoalkane-derived diradicals have all played a part in providing the rich harvest of new bicyclic peroxides described herein.

Not only has the prostanoid [2.2.1] skeleton yielded to the onslaught, but seven other bicyclic peroxide systems have also been obtained. Particularly versatile have been the singlet oxygen-diimide and silver salt routes which have provided higher [n.2.2] systems (n = 2–4), and [n.2.1] systems (n = 3–5) respectively; the [3.3.2] skeleton, unique in its lack of both 5- and 6-membered peroxide rings, was afforded via peroxymercuration.

Doubtless attempts will now be made to prepare more sophisticated model compounds that more closely resemble the natural prostaglandin endoperoxides. Thus the influence of *endo*-5-allyl and/or *exo*-6-vinyl groups upon the chemistry of the [2.2.1] system is clearly of interest. However, it seems likely that the next major effort in this area will involve a thorough and systematic investigation of the reactions of the first generation of bicyclic peroxides that has now been obtained.

5 Acknowledgement

It is a pleasure to acknowledge the skillful and dedicated efforts of our co-workers whose names are to be found among the references and whose achievements have provided the basis of this review. We are also grateful to N.A.T.O. for the provision of Research Grant Number 1559 which has enabled us to collaborate in the production of this article. Our work has been supported by the National Institutes of Health, the National Science Foundation, the Petroleum Research Fund, the Guggenheim Foundation, the Sloan Foundation, Research Corporation, Western Fher Company, Eli Lilly Company, and University of Puerto Rico (WA), and by the Science Research Council (AJB).

6 References

1. Paget, H.: J. Chem. Soc. 829 (1938)
2. von Euler, U. S.: Arch. Exp. Pathol. Pharmakol. *175*, 78 (1934)
3. (a) Bergström, S., Sjövall, J.: Acta Chem. Scand. *14*, 1693 (1960); (b) Bergström, S., Ryhage, R., Samuelsson, B., Sjövall, J.: J. Biol. Chem. *238*, 3555 (1963); (c) Samuelsson, B.: J. Am. Chem. Soc. *85*, 1878 (1963)
4. Samuelsson, B.: J. Am. Chem. Soc. *87*, 3011 (1965)
5. (a) Hamberg, M., Samuelsson, B.: Proc. Nat. Acad. Sci. (USA) *70*, 899 (1973); (b) Nugteren, D. H., Hazelhof, E.: Biochem. Biophys. Acta *326*, 448 (1973)
6. Hamberg, M., Svensson, J., Samuelsson, B.: Proc. Nat. Acad. Sci. (USA) *72*, 2994 (1975)
7. (a) Johnson, R. A. et al.: Prostaglandins *12*, 915 (1976); (b) Corey, E. J., Keck, G. E., Szekély, F.: J. Am. Chem. Soc. *99*, 2006 (1977)
8. Moncada, S., Higgs, E. A., Vane, J. R.: Lancet I, 18 (1977)

9. Hamberg, M., Samuelsson, B.: J. Biol. Chem. *242*, 5336 (1967)
10. Pryor, W. A., Stanley, J. P.: J. Org. Chem. *40*, 3615 (1975)
11. Porter, N. A., Funk, M. O.: ibid. *40*, 3614 (1975)
12. (a) Davies, A. G.: Organic Peroxides, Butterworths, 1961; (b) Swern, D. (ed.): Organic Peroxides, Wiley-Interscience, 1970, vol 1; 1971, vol 2; 1972, vol 3
13. Schenck, G. O., Gollnick, K.: in 1,4-Cycloaddition Reactions (J. Hamer (ed.)), Academic Press, 1967, Chap. 10
14. (a) Hünig, S., Müller, H. R., Thier, W.: Angew. Chem. Int. Ed. Engl. *4*, 271 (1965); (b) Miller, C. E.: J. Chem. Ed. *42*, 254 (1965)
15. Coughlin, D. J., Salomon, R. G.: J. Am. Chem. Soc. *99*, 655 (1977)
16. Adam, W., Eggelte, H. J.: J. Org. Chem. *42*, 3987 (1977)
17. Adam, W., Eggelte, H. J.: unpublished
18. Lightner, D. A., Bisocchi, G. S., Norris, R. D.: J. Am. Chem. Soc. *98*, 802 (1976)
19. (a) Skorianetz, W., Schulte-Elte, K. H., Ohloff, G.: Angew. Chem. Int. Ed. Engl. *11*, 330 (1972); (b) Harada, N., Suzuki, S., Uda, H., Ueno, H.: J. Am. Chem. Soc. *94*, 1777 (1972); (c) Harada, N. et al.: Chem. Lett. 1173 (1973); (d) Harada, N. et al.: ibid. 893 (1974)
20. Adam, W., Erden, I.: Angew. Chem. *90*, 223 (1978)
21. Adam, W., Erden, I.: J. Org. Chem. *43*, 2737 (1978)
22. Takeshita, H., Kanamori, H., Hatsui, T.: Tetrahedron Lett. 3139 (1973)
23. Adam, W., Eggelte, H. J., Rodriguez, A.: Synthesis, 383 (1978)
24. Crigee, R.: Angew. Chem. Int. Ed. Engl. *14*, 745 (1975)
25. Adam, W., Eggelte, H. J.: ibid. *17*, 765 (1978)
26. (a) van Tilborg, W. J. M.: Rec. Trav. Chem. Pays-Bas *95*, 140 (1976); (b) Skold, C. N., Schlessinger, R. H.: Tetrahedron Lett., 791 (1970); (c) Wasserman, H. H., Strehlow, W.: ibid. 795 (1970)
27. Adam, W., Eggelte, H. J.: Angew. Chem. *89*, 762 (1977)
28. Coughlin, D. J., Brown, R. S., Salomon, R. G.: J. Am. Chem. Soc. *101*, 1533 (1979)
29. (a) Adam, W., Erden, I.: Angew. Chem. *90*, 223 (1978); (b) Adam, W., Erden, I.: J. Am. Chem. Soc., *101*, 5692 (1979)
30. (a) Adam, W., Balci, M.: Angew. Chem. Int. Ed. Engl. *17*, 954 (1978); (b) Adam, W., Balci, M.: J. Org. Chem. *44*, 1189 (1979); (c) Adam, W., Balci, M., Pietrzak, B.: J. Am. Chem. Soc. *101*, 6285 (1979); (d) Adam, W., Balci, M.: J. Am. Chem. Soc. *101*, 7537 (1979); (e) Adam, W., Balci, M.: J. Am. Chem. Soc. *101*, 7542 (1979); (f) Adam, W., Balci, M., Pietrzak, B., Rebollo, H.: Synthesis, 820 (1980)
31. (a) Adam, W., Erden, I.: Tetrahedron Lett. *2781* (1979); cf. Errata in Tetrahedron Lett., (1980); (b) Adam, W., Cueto, O., de Lucchi, O.: J. Org. Chem. *45*, 5220 (1980)
32. Adam, W., Balci, M., Rivera, J.: unpublished
33. Adam, W., Erden, I.: Tetrahedron Lett. 1975 (1979)
34. Adam, W. et al.: Angew. Chem. Int. Ed. Engl. *17*, 209 (1978)
35. Cookson, P. G., Davies, A. G., Roberts, B. P.: Chem. Comm. 1022 (1976)
36. Porter, N. A., Gilmore, D. W.: J. Am. Chem. Soc. *99*, 3503 (1977)
37. Davies, A. G., Sotowicz, A. J.: personal communication
38. Bloodworth, A. J., Eggelte, H. J.: Chem. Comm. 741 (1979); J. C. S. Perkin I, 1375 (1981)
39. Bloodworth, A. J., Eggelte, H. J., Galas, A. M. R., Hursthouse, M. B.: unpublished
40. Bloodworth, A. J., Eggelte, H. J.: Tetrahedron Lett. 2001 (1980); 169 (1981)
41. Bloodworth, A. J., Eggelte, H. J.: unpublished
42. Caglioti, L. et al.: Tetrahedron, *34*, 135 (1978)
43. Johnson, R. A. et al.: J. Am. Chem. Soc., *99*, 7738 (1977)
44. Johnson, R. A., Nidy, E. G.: J. Org. Chem. *40*, 1680 (1975)
45. Porter, N. A. et al.: J. Org. Chem. *43*, 2088 (1978)
46. Porter, N. A. et al.: J. Am. Chem. Soc. *101*, 4319 (1979)
47. Porter, N. A. et al.: ibid. *102*, 1183 (1980)
48. Bloodworth, A. J.: in The Chemistry of Mercury (C. A. McAuliffe (ed.)), Macmillan, 1977, Part 3
49. Bloodworth, A. J., Khan, J. A., Loveitt, M. E.: J. C. S. Perkin I, 621 (1981)
50. Bloodworth, A. J., Loveitt, M. E.: ibid. 522 (1978)
51. Bloodworth, A. J., Khan, J. A.: ibid. 2450 (1980)

52. Bloodworth, A. J., Khan, J. A.: Tetrahedron Lett. 3075 (1978)
53. Bloodworth, A. J., Leddy, B. P.: Tetrahedron Lett. 729 (1979)
54. Baldwin, J. E.: Chem. Comm. 734 (1976)
55. Salomon, M. F., Salomon, R. G.: J. Am. Chem. Soc. *101*, 4290 (1979) and references therein
56. Salomon, R. G., Salomon, M. F.: J. Am. Chem. Soc. *99*, 3501 (1977)
57. Wilson, R. M., Geiser, F.: ibid. *100*, 2225 (1978)
58. Adam, W., Bloodworth, A. J.: unpublished
59. Gleiter, R. et al.: J. Electron Spectrosc. Rel. Phen, *19*, 223 (1980)
60. Bloodworth, A. J., Eggelte, H. J., Gleiter, R.: unpublished
61. Zagorski, M. G., Salomon, R. G.: J. Am. Chem. Soc. *102*, 2501 (1980)
62. Salomon, R. G., Salomon, M. F., Coughlin, D. J.: ibid. *100*, 660 (1978)
63. Adam, W.: Acc. Chem. Res. *12*, 390 (1979)
64. (a) Adam, W., Duran, N.: J. Am. Chem. Soc. *99*, 2729 (1977); (b) Richardson, W. H., McGinness, R., and O'Neal, H.: J. Org. Chem. *46*, 1887 (1981)
65. Adam, W., Sanabia, J.: ibid. *99*, 2735 (1977)
66. (a) Hamberg, M., Samuelsson, B.: Proc. Nat. Acad. Sci. (USA) *71*, 3400, 3824 (1974); (b) Diczfalnsy, U., Falardeau, P., Hammarström, S.: FEBS Letters *84*, 271 (1977)
67. Coughlin, D. J., Salomon, R. G.: J. Am. Chem. Soc. *101*, 2761 (1979)
68. Adam, W.: Chem. Unserer Zeit *14*, 44 (1980)

Bei Fragen zur Produktsicherheit wenden Sie sich bitte an:
If you have any questions regarding product safety,
please contact:

Walter de Gruyter GmbH
Genthiner Straße 13
10785 Berlin
productsafety@degruyterbrill.com